T0304016

Safety Performance in a Lean Environment

A GUIDE TO BUILDING SAFETY INTO A PROCESS

Occupational Safety and Health Guide Series

Series Editor

Thomas D. Schneid
Eastern Kentucky University
Richmond, Kentucky

Published Titles

Corporate Safety Compliance: OSHA, Ethics, and the Law
Thomas D. Schneid

Creative Safety Solutions
Thomas D. Schneid

Disaster Management and Preparedness
Thomas D. Schneid and Larry R. Collins

Labor and Employment Issues for the Safety Professional
Thomas D. Schneid

Loss Control Auditing: A Guide for Conducting Fire, Safety, and Security Audits
E. Scott Dunlap

Loss Prevention and Safety Control: Terms and Definitions
Dennis P. Nolan

Managing Workers' Compensation: A Guide to Injury Reduction
and Effective Claim Management
Keith R. Wertz and James J. Bryant

Motor Carrier Safety: A Guide to Regulatory Compliance
E. Scott Dunlap

Occupational Health Guide to Violence in the Workplace
Thomas D. Schneid

Physical Hazards of the Workplace
Larry R. Collins and Thomas D. Schneid

Safety Performance in a Lean Environment: A Guide to
Building Safety into a Process
Paul F. English

Safety Performance in a Lean Environment

A GUIDE TO BUILDING SAFETY INTO A PROCESS

PAUL F. ENGLISH

CRC Press
Taylor & Francis Group
Boca Raton   London   New York

CRC Press is an imprint of the
Taylor & Francis Group, an **informa** business

CRC Press
Taylor & Francis Group
6000 Broken Sound Parkway NW, Suite 300
Boca Raton, FL 33487-2742

Version Date: 20111025

International Standard Book Number: 978-1-4398-2112-1 (Hardback)

Library of Congress Cataloging-in-Publication Data

English, Paul F.
  Safety performance in a lean environment : a guide to building safety into a process / Paul F. English.
       p. cm. -- (Occupational safety & health guide series)
    Includes bibliographical references and index.
    ISBN 978-1-4398-2112-1 (hardback)
    1. Industrial safety. 2. Lean manufacturing. 3. Environmental health. I. Title.

  T55.E45 2011
  658.3'82--dc23   2011038754

Visit the Taylor & Francis Web site at
http://www.taylorandfrancis.com

and the CRC Press Web site at
http://www.crcpress.com

Dedication

This book is dedicated to my wife and children for the unwavering support in my life and career. Each day is a roller coaster that I would never get off. To my wife, Stacy, thank you for all the love and support anyone could possibly have. To my oldest, Madeline, who challenges my wit even at 10 years old. To the greatest negotiator I know, my oldest boy, Donald Callan, who really needs to go into sales when he gets older. To Elizabeth, the quietest one of the bunch, who seems to tolerate all the craziness in our house. To Owen, the youngest, most fearless of all the kids, willing to try anything once and usually does. Thank you and I love you all.

Contents

Foreword

For some people, the combination of lean enterprise and environmental health and safety (EHS) is like mixing oil and water. The reality is that both have many commonalities. In many cases, people focus on lean as only for the manufacturing floor and their processes, but that couldn't be further from the truth. Lean is about behavioral change: a change for the better, identifying simple and waste-free methods to perform any activity, process, or service. In EHS, approximately 85% of all injuries are related to poor behaviors (unsafe acts); therefore we must create a safety culture. EHS professionals must recognize the power of the lean culture and use its power to build upon their own EHS culture. Why not incorporate EHS into the idea of standardized work for all activities, why not incorporate safety into your 5S audits, why not visualize safety within the visual management tools in the workplace, why not make safety part of leadership standard audits and shop floor (Gemba) walks, and why not work with your lean leadership to change the culture together?

EHS already has many of the expected lean tools in its arsenal. When an injury occurs, it is expected that the EHS and leadership team react quickly with a sense of urgency to solve the root cause of the injury. Therefore, we all rush to the area to see what happened (go and see), we investigate and utilize problem-solving techniques (5-Why, 8D, etc.). We do our best to find and eliminate the root cause of that injury and we look for other areas or similar conditions and implement corrective actions (*yokoten* or knowledge sharing). From my 18 years of lean experience, that sure does sound like a strong lean culture and methodology. The expectation of lean is to identify waste and eliminate that waste from the process so it never returns and, in the spirit of continuous improvement (*kaizen*), to constantly be looking for additional opportunities to improve on a minute-by-minute basis. That is why we use tools like visual management to highlight or identify the abnormality so we can react to it with a sense of urgency and solve it for root cause; we then standardize the process or activity to ensure that it is sustained and finally share the improvements with other areas or groups so they too can gain from our experience. So, EHS and lean should share the same systems and methods to solve these issues identified in both our proactive and reactionary methods.

Last, EHS and lean must work together as we both have one common enemy: sustainment. EHS programs as well as lean programs need continuous monitoring and consistency to sustain the cultures that have been built. This relationship is symbiotic; if a lean process is compromised, frequency and severity will increase for injuries. Thus, lean and EHS must work together to define a strong layered audit structure defined in a leadership standardized work process to repetitively drive the expected behavioral changes. This type of audit process is the key to sustainment and true culture change in both the EHS and lean processes.

David Moore
Lean Director, E-ONE, Inc.

Acknowledgments

Special thanks to Peter Guile, president of E-ONE, Inc., and the E-ONE team for helping me write this book. All the determination, unwavering focus, and hard work have made it possible to achieve a world-class lean enterprise over the past few years. This change not only demonstrates the unmatched quality of the fire trucks produced by E-ONE but also, more impressive, the transformation and expectations of E-ONE team members. A culture of pride and employee engagement has been created and harnessed, making extraordinary results possible. E-ONE has pursued a lean transformation with a focus on customer voice in design, performance, and crash protection. Participating in this experience while championing the EHS initiatives for E-ONE has been invaluable in the writing of this book.

Today's fire service does so much more than "put wet stuff on the red stuff"; firefighters are true *first responders*. Many are volunteers who selflessly provide medical, rescue, natural disaster, and firefighting services to their communities. To be a member of the E-ONE team is to serve our nation's heroes by designing and building the best custom mission-critical vehicles possible. E-ONE recognizes the privilege and honor to be a part of the fire service.

All photographs in this book are used with permission and at the courtesy of E-ONE, Inc. Web downloads for this book are available at http://www.crcpress.com/product/isbn/9781439821121

The Author

Paul English is an assistant professor at Eastern Kentucky University in the Safety, Security & Emergency Management program. He has worked for Fortune 100 companies including Nestlé and Ford Motor Company in different facets of occupational safety, security, and emergency response operations. While at Ford Motor Company, he was the recipient of the President's Health & Safety Award for Innovation representing the Americas. His most recent previous position was Director of Environmental Health and Safety with E-ONE, Inc., in Ocala, Florida.

He earned a B.S. in fire and safety engineering technology, specializing in industrial risk management, and an M.S. safety, security, and emergency management, both from Eastern Kentucky University. He has numerous publications regarding incident investigation, process safety management, benchmarking safety metrics, and emergency response. He is an active member of the American Society of Safety Engineers and is a Certified Safety Professional (CSP).

1 Management Models and Lean Processes

Failure is simply the opportunity to begin again, this time more intelligently.

—Henry Ford

Since the dawn of the industrial revolution, many different management models have been implemented, tested, and improved. A look at different management models from a lean perspective allows us to determine what model fits the best when providing a finished good or service. In lean enterprise, Henry Ford is considered by many to be the grandfather of lean management and thinking, mostly for the simple reason of his ability to mass-produce a finished good. Ford's mistakes were opportunities to provide what was missing in the product, variation, and options. By no means is this section of the book a comprehensive history lesson in modern management theory, but rather a quick snapshot of how lean enterprise and management came to be in the modern era.

HENRY FORD AND THE FORD MOTOR COMPANY, MASS PRODUCTION

Henry Ford, in his early career, worked at Westinghouse as a maintenance employee servicing steam engines. Most of his knowledge of steam engines came to him while working on the family farm. It was during this time that Ford came to the conclusion that the steam engine was not suitable for light vehicles, mostly due to weight issues (Ford and Crowther 1923). After this self-realization, Ford figured that there was nothing left for him to do or learn from steam engines and left Westinghouse after only one year. He then went to work for Detroit Electric Company and was appointed an engineer. It was during this time that he built and sold several gas-powered cars, locally in Detroit. Ford was seen as a visionary at the electric company and was asked to give up on the gas-powered engine, as electricity was seen as the wave of the future. Ford quit his job in 1899 and went into the automobile business full-time. Between 1902 and 1908, Ford started and left the Detroit Automobile Company, which would later become Cadillac. In 1903, the Ford Motor Company was formed a week after one of Ford's cars won a race in Michigan.

In 1908, Ford introduced the Model T, which was an instant hit. Ford had identified the customer and what the costumer wanted, which was a low-cost vehicle that was easy to operate and cheap to repair if damaged. The Model T was so popular that at one time Ford Motor Company held 50% of the market share for all automobiles in the United States. Ford in his autobiography stated,

In the success of the Ford car the early provision of service was an outstanding element. Most of the expensive cars of that period were ill provided with service stations. If your car broke down you had to depend on the local repair man—when you were entitled to depend upon the manufacturer. If the local repair man were a forehanded sort of a person, keeping on hand a good stock of parts (although on many of the cars the parts were not interchangeable), the owner was lucky. But if the repair man were a shiftless person, with an adequate knowledge of automobiles and an inordinate desire to make a good thing out of every car that came into his place for repairs, then even a slight breakdown meant weeks of laying up and a whopping big repair bill that had to be paid before the car could be taken away. The repair men were for a time the largest menace to the automobile industry. (Ford and Crowther 1923)

In 1913, Ford introduced the concept of the moving assembly line, which decreased the cost to produce vehicles and increased productivity. The moving assembly line that Ford conceived did have a drawback, as all vehicles were painted black due to the drying time requirements of the paint. Ford also eliminated any type of variation in the Model T to reduce cost. In turn, this meant that little to no options were available in the Model T, which opened the door for aftermarket ideas and suppliers as well as the competition.

Management decisions were made directly by Henry Ford until he semi-retired, leaving all decision-making authority to his grandson in 1945. In his day, Ford was also very stubborn, which many believe was to blame for the inflexibility of the company. Ford's inability to change with the marketplace and customer need opened the door to competitors. To many, Henry Ford is considered the grandfather of lean enterprise. His contributions include, but are not limited to, the following:

- Identifying the voice of the customer (quality product, easy to use, cheap to service)
- Identifying and eliminating waste in the manufacturing process
- Creating the moving assembly line (introducing process flow)
- Creation of the 8-hour work shift (only way to evenly distribute time for a 24-hour operation)
- Supplier quality (Ford believed in total supplier control to ensure quality. Many assembly plants were fully integrated, meaning that all parts were built by Ford. The Rouge Complex included a fully integrated steel mill where Ford could control the quality of steel being produced to build his vehicles.)

Henry Ford was a true visionary and industrial pioneer whose legacy still lives on today in the Ford Motor Company.

ALFRED SLOAN AND GENERAL MOTORS CORPORATION, MASS PRODUCTION—HIGH VARIATION

Alfred Sloan was the owner of Hyatt Roller Bearings, which was a supplier to several automobile manufacturers at the time. Hyatt was merged with several other companies and later came to be known as General Motors (GM). Sloan became vice president and then president of General Motors in 1923.

Sloan was able to identify weaknesses in Henry Ford's inability and unwillingness to provide any options for prospective customers. Sloan created plans to make automobile models obsolete, thus creating model changes after several years. To increase sales, Sloan offered GM vehicles with a variety of different colors with the help of GM's largest stockholder at the time, DuPont. The business relationship of DuPont supplying automotive-grade paint to GM is still in place today. These ideas pushed GM past Ford in sales as the Model T languished from lack of change and options.

Sloan is also considered the creator of the return on investment (ROI) model as well as the accounting process of counting physical inventory of finished goods as a cash value. The marketing of GM products while Sloan was president came to be known as the "ladder of success." GM allowed customers to "climb the ladder of success" by trading in old vehicles for new, more affluent models. The ladder was bottom to top, Chevrolet, Pontiac, Oldsmobile, Buick, and Cadillac. This business model was emulated by many other automobile manufacturers with the addition of different nameplates within the same company. During Sloan's reign at GM, the financial arm of GM was created. GMAC offered customers an avenue to purchase a GM vehicle while making incremental monthly payments.

Alfred Sloan and his work at GM created what Ford lacked, variety and options. GM was able to capitalize on the mass production ideas of Ford, while adding different options that customers were willing to pay for.

EDWARD DEMING AND TOTAL QUALITY MANAGEMENT

Dr. William Edwards Deming was a statistician born in the United States in 1900 who went on to find huge success in Japan after World War II. Deming taught senior management in Japan how to improve quality through different methods, including the use of statistical analysis of products. Deming is considered to have had the greatest impact on Japanese manufacturing and management models without having any Japanese heritage. His work went largely unnoticed in the United States until the early 1980s when Japanese automobile manufacturing overtook GM, Ford, and Chrysler as the leaders in quality and production.

In 1981, Ford Motor Company went to Deming for help to improve the quality of vehicle lines. Once all the issues were reviewed, he determined that the quality system was not at fault for Ford's woes. He insisted that the management practices were a direct cause of the quality issues and that a cultural change needed to take place to build quality into the process of building automobiles. In 1985, Ford launched one of the most successful vehicles in history, the Taurus/Sable platform.

Dr. Deming has authored several books on quality and management theories that are still taught at different levels in a variety of organizations around the world. His ideas of total quality management not only in business but also in life and philosophy have been handed down from generation to generation in the Japanese culture. The philosophy piece of Deming's teachings is the sole reason why few if any companies will reach the level of lean thinking that Toyota has attained. The Toyota Production System (TPS) is the model that many companies choose to implement and emulate lean enterprise in an organization. Principle 1 of the Toyota Way is to "base your management decisions on a long term philosophy, even at the cost of short term financial goals" (Liker 2004). Toyota's development of hybrid technology and subsequent launch of the Prius is a prime example of Deming's cultural influence. Deming developed the system of profound knowledge, which is the basis of his 14 points of management.

THE DEMING SYSTEM OF PROFOUND KNOWLEDGE

The prevailing style of management must undergo transformation. A system cannot understand itself. The transformation requires a view from outside. The aim of this chapter is to provide an outside view—a lens—that I call a system of profound knowledge. It provides a map of theory by which to understand the organizations that we work in.

The first step is transformation of the individual. This transformation is discontinuous. It comes from an understanding of the system of profound knowledge. The individual, transformed, will perceive new meaning to his life, to events, to numbers, to interactions between people.

Once the individual understands the system of profound knowledge, he will apply its principles in every kind of relationship with other people. He will have a basis for judgment of his own decisions and for transformation of the organizations that he belongs to. The individual, once transformed, will do the following:

- Set an example
- Be a good listener, but will not compromise
- Continually teach other people
- Help people to pull away from their current practice and beliefs and move into the new philosophy without a feeling of guilt about the past

The layout of profound knowledge appears here in four parts, all related to each other:

- Appreciation for a system
- Knowledge about variation
- Theory of knowledge
- Psychology (Deming Institute 2011)

ORIGIN OF THE 14 POINTS

The 14 points (see Figure 1.1) are the basis for the transformation of American industry. It will not suffice merely to solve problems, big or little. Adoption and action on the 14 points are a signal that the management intends to stay in business and aims to protect investors and jobs. The system Deming formed was the basis for management models for Japan in 1950 and the years that followed. The 14 points apply anywhere, to small organizations as well as to large ones, to the service industry as well as to manufacturing. They apply to a division within a company.

Deming's 14 points demonstrate his position that a business that fails to stay innovative and plan for the future will lose market share and, in turn, lose people. He stressed that a company must invest in people and long-term learning to keep ahead of the competition. After the release of his book *Out of the Crisis* in 1986, he was credited by many as the father of the Total Quality Management (TQM) model.

1. Create constancy of purpose toward improvement of product and service, with the aim to become competitive and to stay in business, and to provide jobs.
2. Adopt the new philosophy. We are in a new economic age. Western management must awaken to the challenge, learn their responsibilities, and take on leadership for change.
3. Cease dependence on inspection to achieve quality. Eliminate the need for inspection on a mass basis by building quality into the product in the first place.
4. End the practice of awarding business on the basis of price tag. Instead, minimize total cost. Move toward a single supplier for any one item, on a long-term relationship of loyalty and trust.
5. Improve constantly and forever the system of production and service, to improve quality and productivity, and thus constantly decrease costs.
6. Institute training on the job.
7. Institute leadership. The aim of supervision should be to help people, machines, and gadgets to do a better job. Supervision of management is in need of overhaul, as well as supervision of production workers.
8. Drive out fear, so that everyone may work effectively for the company.
9. Break down barriers between departments. People in research, design, sales, and production must work as a team, to foresee problems of production and in use that may be encountered with the product or service.
10. Eliminate slogans, exhortations, and targets for the workforce asking for zero defects and new levels of productivity. Such exhortations only create adversarial relationships, as the bulk of the causes of low quality and low productivity belong to the system and thus lie beyond the power of the workforce.
11. Eliminate work standards (quotas) on the factory floor. Substitute leadership. Eliminate management by objective. Eliminate management by numbers, numerical goals. Substitute leadership.
12. Remove barriers that rob the hourly worker of his right to pride of workmanship. The responsibility of supervisors must be changed from sheer numbers to quality. Remove barriers that rob people in management and in engineering of their right to pride of workmanship. This means, inter alia, abolishment of the annual or merit rating and of management by objective.
13. Institute a vigorous program of education and self-improvement.
14. Put everybody in the company to work to accomplish the transformation. The transformation is everybody's job.

FIGURE 1.1 Deming's 14 points. (From Deming, W. Edwards. "Out of the Crisis." In *Out of the Crisis*, by W. Edwards Deming, 23–24. Cambridge, MA: MIT Press, 2000. © 2000 Massachusetts Institute of Technology, by permission of MIT Press.)

EIJI TOYODA AND TOYOTA PRODUCTION SYSTEM, TPS—MODERN-DAY LEAN MANAGEMENT

After World War II, several members of the Toyoda family visited the United States at the behest of Dr. Deming. Many in the postwar Japanese culture drew from Deming's research on quality control and business operations. Ford Motor Company was on the list to benchmark what was at that time the most powerful and profitable company in the world. Upon benchmarking several facilities, it was determined that large amounts of excess inventory and excessive amounts of waste proved that the business model could be drastically improved.

During the trip, the group went to a supermarket and was surprised at what they found. The level of visual management and replacement of sold inventory in the store lent itself to the waste found at the Ford facilities. Toyoda coupled the assembly line idea with a replenishment system that delivered parts Just-in-Time (JIT) to build products more efficiently.

Between 1948 and 1975, Toyota developed what is today called the Toyota Production System (TPS). Over recent years, the system has begun to be called the Thinking People System, as many companies have implemented lean enterprise modeled after Toyota in non-automobile businesses. Taiichi Ohno, Shigeo Shingo, and Eiji Toyoda are considered the fathers of TPS. TPS has been written about for years in different contexts, and in 2004 Dr. Jeffrey Liker published *The Toyota Way*, which is widely acclaimed to be the catchall book on Toyota's system and business philosophy. (See Figure 1.2.)

Dr. Liker interviewed countless Toyota employees and suppliers to develop what TPS is, how it works, and, more importantly, why it works so well. Liker came up with four broad categories to define TPS with 14 supporting principles behind the categories.

As discussed earlier, TPS has been infused into the culture of Toyota in such a way that it penetrates all levels of the organization from the top down. Dr. Liker dedicated a chapter in *The Toyota Way* regarding the forward thinking of the company in the development of the Prius. The Prius began its life in 1993, starting in development and launching 2 months ahead of schedule in 1997. Although sales of the car exceeded expectations, it is still unclear if Toyota has seen any profit on the Prius, as it enters its third generation. Toyota based the decision to develop and manufacture the vehicle on a long-term philosophy at the expense of profit and short-term financial goals. This is just one example of why other companies fall short of trying to catch Toyota as a company. How many managers or CEOs can go before a superior or board of directors and say, "We are going to build a product and spend over 1 billion dollars to develop and manufacture it. We will not see a profit for this product in the next 5–6 years"?

MANAGEMENT PROCESSES

As many people move through their careers and work for different companies, many management styles, processes, and systems can be identified. What many companies have done as part of the overall continuous improvement process is to identify different aspects of management processes for different applications. Different management processes lend themselves better to different industries.

Management by objectives (MBO) is the process of managing an organization through levels of cascading objectives. Employees are given individual objectives that will roll into a specific department objective and finally into company or organizational objectives. In some organizations, employees will select their own objectives as long as they correlate with the goals of the department, facility, or company. This example given is a very simple model, but in large organizations MBO can create very complex models and requires extensive feedback to ensure that goals are being achieved. Process checks usually occur once a quarter or biannually, depending on the goals or objectives that have been identified by the management team. At

SECTION I: LONG-TERM PHILOSOPHY

Principle 1—Base your management decisions on a long-term philosophy, even at the expense of short-term financial goals.

- Have a philosophical sense of purpose that supersedes any short-term decision making. Work, grow, and align the whole organization toward a common purpose that is bigger than making money. Understand your place in the history of the company and work to bring the company to the next level. Your philosophical mission is the foundation for all the other principles.
- Generate value for the customer, society, and the economy—it is your starting point. Evaluate every function in the company in terms of its ability to achieve this.
- Be responsible. Strive to decide your own fate. Act with self-reliance and trust in your own abilities. Accept responsibility for your conduct and maintain and improve the skills that enable you to produce added value.

SECTION II: THE RIGHT PROCESS WILL PRODUCE THE RIGHT RESULTS

Principle 2—Create a continuous process flow to bring problems to the surface.

- Redesign work processes to achieve high value-added, continuous flow. Strive to cut back to zero the amount of time that any work project is sitting idle or waiting for someone to work on it.
- Create flow to move material and information fast as well as to link processes and people together so that problems surface right away.
- Make flow evident throughout your organizational culture. It is the key to a true continuous improvement process and to developing people.

Principle 3—Use "pull" systems to avoid overproduction.

- Provide your down line customers in the production process with what they want, when they want it, and in the amount they want. Material replenishment initiated by consumption is the basic principle of Just-in-Time.
- Minimize your work in process and warehousing of inventory by stocking small amounts of each product and frequently restocking based on what the customer actually takes away.
- Be responsive to the day-by-day shifts in customer demand rather than relying on computer schedules and systems to track wasteful inventory.

FIGURE 1.2 Executive Summary of the 14 Toyota Way Principles. (From Liker, Jeffrey K. *The Toyota Way*. New York: McGraw-Hill, 2004. © 2004, The McGraw-Hill Companies, reprinted with permission.)

Principle 4—Level out the workload (*heijunka*). (Work like the tortoise, not the hare.)

- Eliminating waste is just one-third of the equation for making lean successful. Eliminating overburden to people and equipment and eliminating unevenness in the production schedule are just as important—yet generally not understood at companies attempting to implement lean principles.
- Work to level out the workload of all manufacturing and service processes as an alternative to the stop/start approach of working on projects in batches that is typical at most companies.

Principle 5—Build a culture of stopping to fix problems, to get quality right the first time.

- Quality for the customer drives your value proposition.
- Use all the modern quality assurance methods available.
- Build into your equipment the capability of detecting problems and stopping itself. Develop a visual system to alert team or project leaders that a machine or process needs assistance. *Jidoka* (machines with human intelligence) is the foundation for "building in" quality.
- Build into your organization support systems to quickly solve problems and put in place countermeasures.
- Build into your culture the philosophy of stopping or slowing down to get quality right the first time to enhance productivity in the long run.

Principle 6—Standardized tasks and processes are the foundation for continuous improvement and employee empowerment.

- Use stable, repeatable methods everywhere to maintain the predictability, regular timing, and regular output of your processes. It is the foundation for flow and pull.
- Capture the accumulated learning about a process up to a point in time by standardizing today's best practices. Allow creative and individual expression to improve upon the standard; then incorporate it into the new standard so that when a person moves on you can hand off the learning to the next person.

Principle 7—Use visual control so no problems are hidden.

- Use simple visual indicators to help people determine immediately whether they are in a standard condition or deviating from it.
- Avoid using a computer screen when it moves the worker's focus away from the workplace.

FIGURE 1.2 *(Continued)*

- Design simple visual systems at the place where the work is done, to support flow and pull.
- Reduce your reports to one piece of paper whenever possible, even for your most important financial decisions.

Principle 8—Use only reliable, thoroughly tested technology that serves your people and processes.

- Use technology to support people, not to replace people. Often it is best to work out a process manually before adding technology to support the process.
- New technology is often unreliable and difficult to standardize and therefore endangers "flow." A proven process that works generally takes precedence over new and untested technology.
- Conduct actual tests before adopting new technology in business processes, manufacturing systems, or products.
- Reject or modify technologies that conflict with your culture or that might disrupt stability, reliability, and predictability.
- Nevertheless, encourage your people to consider new technologies when looking into new approaches to work. Quickly implement a thoroughly considered technology if it has been proven in trials and it can improve flow in your processes.

SECTION III: ADD VALUE TO THE ORGANIZATION BY DEVELOPING YOUR PEOPLE

Principle 9—Grow leaders who thoroughly understand the work, live the philosophy, and teach it to others.

- Grow leaders from within, rather than buying them from outside the organization.
- Do not view the leader's job as simply accomplishing tasks and having good people skills. Leaders must be role models of the company's philosophy and way of doing business.
- A good leader must understand the daily work in great detail so he or she can be the best teacher of your company's philosophy.

Principle 10—Develop exceptional people and teams who follow your company's philosophy.

- Create a strong, stable culture in which company values and beliefs are widely shared and lived out over a period of many years.
- Train exceptional individuals and teams to work within the corporate philosophy to achieve exceptional results. Work very hard to reinforce the culture continually.

FIGURE 1.2 *(Continued)*

- Use cross-functional teams to improve quality and productivity and enhance flow by solving difficult technical problems. Empowerment occurs when people use the company's tools to improve the company.
- Make an ongoing effort to teach individuals how to work together as teams toward common goals. Teamwork is something that has to be learned.

Principle 11—Respect your extended network of partners and suppliers by challenging them and helping them improve.

- Have respect for your partners and suppliers and treat them as an extension of your business.
- Challenge your outside business partners to grow and develop. It shows that you value them. Set challenging targets and assist your partners in achieving them.

SECTION IV: CONTINUOUSLY SOLVING ROOT PROBLEMS DRIVES ORGANIZATIONAL LEARNING

Principle 12—Go and see for yourself to thoroughly understand the situation (*genchi genbutsu*).

- Solve problems and improve processes by going to the source and personally observing and verifying data rather than theorizing on the basis of what other people or the computer screen tell you.
- Think and speak based on personally verified data.
- Even high-level managers and executives should go and see things for themselves, so they will have more than a superficial understanding of the situation.

Principle 13—Make decisions slowly by consensus, thoroughly considering all options; implement decisions rapidly (*nemawashi*).

- Do not pick a single direction and go down that one path until you have thoroughly considered alternatives. When you have picked, move quickly and continuously down the path.
- *Nemawashi* is the process of discussing problems and potential solutions with all of those affected, to collect their ideas and get agreement on a path forward. This consensus process, though time-consuming, helps broaden the search for solutions, and once a decision is made, the stage is set for rapid implementation.

FIGURE 1.2 *(Continued)*

Principle 14—Become a learning organization through relentless reflection (*hansei*) and continuous improvement (*kaizen*).

- Once you have established a stable process, use continuous improvement tools to determine the root cause of inefficiencies and apply effective countermeasures.
- Design processes that require almost no inventory. This will make wasted time and resources visible for all to see. Once waste is exposed, have employees use a continuous improvement process (*kaizen*) to eliminate it.
- Protect the organizational knowledge base by developing stable personnel, slow promotion, and very careful succession systems.
- Use *hansei* (reflection) at key milestones and after you finish a project to openly identify all the shortcomings of the project. Develop countermeasures to avoid the same mistakes again.
- Learn by standardizing the best practices, rather than reinventing the wheel with each new project and each new manager.

FIGURE 1.2 *(Continued)*

the same time, goals and objectives should be reviewed to ensure that the scope of work is on target and not creeping into unwanted areas.

It was primarily for this reason that Deming disagreed with the MBO thought process. Deming argued that the setting of goals by objectives ignored certain tools and systems in a process, which would lead to a failure to achieve the objective. Giving employees the power to select their own objectives has been shown to empower employees to achieve the goals or objectives identified. This thought process has also led to the development of SMART objectives or goals. SMART stands for *specific, measurable, achievable, realistic*, and *time-based*. All goals and objectives identified as part of MBO philosophy should have all of these attributes reviewed when setting goals.

- Specific: All goals and objectives need to be as specific as possible. General statements need to be avoided when setting specific needs. Statements such as "Reduce incidents and injuries" should be replaced with "Reduce sprains and strains by six incidents or 30%."
- Measurable: Employees and managers must ask themselves, "How will this goal or objective be measured?" You need to understand if you are winning or losing the battle to achieve.
- Achievable: At this point, you should have identified a goal or objective that is specific in nature and can be measured in some way. The next question that needs to be asked is, "Can I achieve this goal?" In many organizations, the goal of zero incidents or injuries is always set as the ideal state. Is this achievable? Some say yes; others say no. The question is, would you sign yourself up for this objective?

- Realistic: All goals and objectives need to recognize that there are variables that can't be controlled or sometimes identified when planning. Realistic characteristics should identify that the entire goal or objective may not be achieved and that success might be small. At the same time, people should be challenged with tasks that are in fact realistic.
- Time-based: Simply stated, "Can the goal be achieved in the amount of time allowed or expected?" As with any plan, priorities can change. If time becomes an issue, more resources can be identified and allocated to ensure that the goal is met on time. That is also why MBO drives periodic checks in the progress to achievement.

SMART objectives can be identified at every level of an organization and can tie individual goals and cascade up to the overall business goals when done correctly. Like any other management model, failure of support and buy-in at the top management levels will most likely lead to failure of the MBO model due to lack of accountability.

BALANCED SCORECARDS

Balanced scorecards (BSC) are another management process or tool used by many organizations to drive performance. Often considered a strategic performance management tool, balanced scorecards try to identify areas in the management process that are unbalanced or subpar. Many different models of BSC have emerged over the past several years. Again, different processes and models require different tools for measurement. The overall implementation of BSC is the same, with different metrics.

BSC were developed by Arthur M. Schneiderman in the late 1980s while working for a company called Analog Devices. Schneiderman states that during his tenure at Analog from 1986 through 1992, he was the process owner for nonfinancial performance measurement and the BSC. In the process of building the BSC, the main objective was to "demonstrate how Analog linked its performance measurement system to its Corporate Objective and business strategy and the richness and completeness of the improvement infrastructure that we had in place" (Schneiderman 2006).

BSC are like any other tool in business management. If the tool is value added to the customer, which in this case would be management, the tool will be used. Where many scorecards fail is the fact that it is looked upon as one more piece of paper to manage or process to update. As BSC looks at bigger and bigger pictures, the process of updating information and management of the process should become smaller and smaller. In a nonfinancial situation in dealing with internal factors such as manufacturing, BSC can become very cumbersome to understand as well as update.

In the original BSC model identified by Schneiderman and later documented by Robert S. Kaplan and David P. Norton in *Using the Balanced Scorecard as a Strategic Management System*, four specific areas were identified to create a BSC.

THE LEARNING AND GROWTH PERSPECTIVE

This perspective includes employee training and corporate cultural attitudes related to both individual and corporate self-improvement. In a knowledge-worker organization, people—the only repository of knowledge—are the main resource. In the current climate of rapid technological change, it is becoming necessary for knowledge workers to be in a continuous learning mode. Metrics can be put into place to guide managers in focusing training funds where they can help the most. In any case, learning and growth constitute the essential foundation for success of any knowledge-worker organization.

Kaplan and Norton emphasize that *learning* is more than *training*; it also includes things like mentors and tutors within the organization, as well as that ease of communication among workers that allows them to readily get help on a problem when it is needed. It also includes technological tools—what the Baldrige criteria call "high-performance work systems."

THE BUSINESS PROCESS PERSPECTIVE

This perspective refers to internal business processes. Metrics based on this perspective allow the managers to know how well their business is running, and whether its products and services conform to customer requirements (the mission). These metrics have to be carefully designed by those who know these processes most intimately; with our unique missions these are not something that can be developed by outside consultants.

THE CUSTOMER PERSPECTIVE

Recent management philosophy has shown an increasing realization of the importance of customer focus and customer satisfaction in any business. These are leading indicators: if customers are not satisfied, they will eventually find other suppliers that will meet their needs. Poor performance from this perspective is thus a leading indicator of future decline, even though the current financial picture may look good. In developing metrics for satisfaction, customers should be analyzed in terms of kinds of customers and the kinds of processes for which we are providing a product or service to those customer groups.

THE FINANCIAL PERSPECTIVE

Kaplan and Norton do not disregard the traditional need for financial data. Timely and accurate funding data will always be a priority, and managers will do whatever is necessary to provide it. In fact, often there is more than enough handling and processing of financial data. With the implementation of a corporate database, it is hoped that more of the processing can be centralized and automated. But the point is that the current emphasis on financials leads to the "unbalanced" situation with regard to other perspectives. There is perhaps a need to include additional financial-related data, such as risk assessment and cost-benefit data, in this category (Institute 1998–2011).

If a BSC is to be used to help manage a business or operations, it must provide and measure data that is being driven from the top down and to all levels of the organization.

BENCHMARKING

Benchmarking is a management tool used to identify best practices in a given setting. The setting can be within a large company with multiple locations, or it can occur within an industry. In the early days of the industrial revolution, benchmarking was unheard of. Many different manufacturing processes were being developed and implemented at the same time. The thought of inviting the competition into your operation would have been viewed as corporate spying or espionage. With the exception of a few industries, benchmarking has become a way for different businesses to learn from one another and help foster continuous improvement of different processes.

Benchmarking processes at different locations and companies can stem from any number of different problems. Safety has always been a good benchmarking platform for many companies to identify best, safe work practices. This is in part due to the fact that many OSHA regulations are uniform to almost all processes and industries. In other words, OSHA 1910.146, which is the confined space standard, will require the same standard to be met in an auto assembly plant or a food manufacturing plant. What can be different is how the location, company, or industry approaches and complies with the safety standard. Because the national standards set by OSHA and EPA are generic and based on applicability, benchmarking for safety is generally a widely accepted practice. The idea to find out how to do something safer, while reducing injuries and illnesses, is hard to pass up.

SPAN OF CONTROL

Span of control refers to the number of employees a supervisor or manager can control to be effective as a manager. Many experts have placed the number of direct reports at a maximum of six employees to maintain close control (Robbins and DeCenzo 2005). As highly desirable as this model is, it is the ideal state in most companies. The current state has some supervisors and managers having control of 30–40 employees directly reporting to them.

Some reasons why a span of control can increase will depend on several different variables. A highly trained and competent workforce or workgroup can function and make decisions on its own, relying on management only as a type of in-house consultant to help with decision making from time to time. Again, standardized workplaces and documented standardized work will allow for a bigger span of control if needed.

Another reason why span of control would be increased is the ability to make decisions faster. In the creation of the U.S. Department of Homeland Security (DHS) the Federal Emergency Management Agency (FEMA) was moved under the control of DHS. The thought process behind moving this agency under DHS control was that it would allow for quicker decisions, better needs analysis, and, most important, a bigger budget. What had turned out to be a positive move to increase a response capability became tangled in bureaucratic red tape and oversight. The response to Hurricane Katrina in 2005 reinforced the idea that increasing a government span of control can lead to negative effects to say the least. Bigger is not always better.

SOCIAL MANAGEMENT

The word *social*, by definition, refers to the decision to live within a community rather than live alone. When we talk about social management in the context of business, we are talking about how a company or organization operates within the public eye or community. Today, many companies devote entire teams and resources to help create a social value for an organization. Many companies have formed endowments and give gifts to various nonprofit organizations, such as the United Way and the Juvenile Diabetes Foundation. The creation of these systems is to give back to the community as a sense of social responsibility for being in business.

Many companies form their business ethics under the social management umbrella. Creation of business ethics helps ensure that all social obligations are being met by all employees. In many organizations, a clear violation of ethics is considered grounds for termination. Depending on what ethics were violated, bad press surrounding poor social management can damage a company's image and profits. Social management has been called into question after such high-profile incidents as the *Exxon Valdez* oil spill in 1989 and the BP Deep Water Horizon incident in 2010. In both cases, social backlash led to stock prices dropping almost immediately as investors quickly tried to put as much distance between themselves and the companies. In addition, consumers also joined in the outrage at lack of social management by both companies by forming boycotts of products and services. BP had significant social management issues that were amplified by Tony Hayward, CEO of BP at the time of the Deepwater Horizon incident. A string of interview miscues and poor word choices led to the removal of Hayward from his position.

Poor social management from an environmental health and safety standpoint can have a severe impact on an organization. EHS professionals always need to be diligent in all aspects of their jobs.

VALUE-BASED MANAGEMENT

Value-based management (VBM) means that a company or organization runs to maximize value or maximize shareholder value. The idea behind this management philosophy is to create a deeper meaning for an organization to return value to shareholders. Organizations that run a VBM system will usually have some type of stock option for an employee to become part of the value-based system. The general idea is that the employee will have a vested interest in the success or failure of the company and, in turn, will create a sense of urgency in the workplace.

There are three main factors that drive VBM:

1. Creating value, both present state and future state
2. Managing for value through:
   a. Governance
   b. Change management
   c. Culture
   d. Communication
   e. Leadership
3. Measuring value

Value-based management is extremely dependent on an organization's values and purpose as the driving force behind the management system (Value Based Management. Net 2011). Another way of looking at VBM metrics, according to the Center for Economic and Social Justice, is that VBM balances moral values along with material values:

1. A foundation of **shared ethical values**—starting with the belief in the intrinsic value of each person (each employee, customer, and supplier)
2. Success in the marketplace based on **delivering maximum value** (higher quality at lower prices) to the customer
3. **Rewards based on the value people contribute** to the company, as individuals and as a team, as workers and as owners

These beliefs and values are periodically checked against norms and adjusted correctly (Center for Economic and Social Justice).

There are many different management models that can be used when a company goes lean. It is up to management to decide how an organization will achieve its "ideal state" of business. As with any program or possible philosophy, whether it be lean enterprise or something entirely different, sustaining is the hardest issue to overcome. Business management models and processes will only be sustained with continued and relentless consistency. As EHS professionals, being consistent every time is what drives results and eliminates variation in the EHS program.

WORKS CITED

Center for Economic and Social Justice. *Value Based Management: A System for Transforming the Corporate Culture.* http://cog.kent.edu/lib/cesj1.htm (accessed January 7, 2011).

Deming, W. Edwards. "Out of the Crisis." In *Out of the Crisis*, by W. Edwards Deming, 23–24. Cambridge, MA: MIT Press, 2000.

Deming Institute. *Teachings.* 2011. http://deming.org/index.cfm?content=66 (accessed January 21, 2011).

Ford, Henry, and Samuel Crowther. *My Life and Work.* Garden City, NY: Country Life Press, 1923.

Liker, Jeffrey K. *The Toyota Way.* New York: McGraw-Hill, 2004.

Robbins, Stephen P., and David A. DeCenzo. "Basic Organization Designs." In *Fundamentals of Management*, 163. Upper Saddle River, NJ: Pearson Prentice Hall, 2005.

Schneiderman, Arthur M. *Analog Devices: 1986–1991—The First Balanced Scorecard.* August 13, 2006. http://www.schneiderman.com (accessed March 29, 2011).

Value Based Management.Net. *What Is Value Based Management?* January 7, 2011. http://www.valuebasedmanagement.net/faq_what_is_value_based_management.html (accessed March 4, 2011).

Wikipedia. *Henry Ford.* July 8, 2009. http://en.wikipedia.org/wiki/Henry_Ford (accessed January 14, 2011).

2 Planning, Decision Making, and Problem Solving

A relentless barrage of "why's" is the best way to prepare your mind to pierce the clouded veil of thinking caused by the status quo. Use it often.

—Shigeo Shingo

STRATEGIC AND TACTICAL PLANNING

Strategic planning often refers to what the overall goal is for a department, organization, or company. Strategic planning can include a company's plan to increase market share, enter a new market, or introduce a new product. In strategic planning, the plan asks the generic question, "Where does the company want to be next year, in 3 years, in 5 years?" In many cases of strategic planning, companies will identify specific goals and measurements including identifying the company's visions, missions, and values.

A vision is a view of what the company wants to look like in its "future or ideal state." This is a long-term view of the company's or organization's position in the marketplace. Depending on what vision is created, it can also be viewed as the "ideal state" of the company. This means that you live in a perfect world, where time, money, and engineering have no bounds.

A mission or mission statement is what the company or organization does to drive to the vision identified. Values are the beliefs of the company or organization that are shared with all employees and shareholders. Without values identified in the strategic planning process, there is no culture or priorities for the company to make decisions to move the business forward. Tactical planning refers to the planning process of how the company or organization will fulfill its strategic plan. Tactical plans often refer to how the day-to-day operations are carried out in an organization. All tactical plans should be tied directly to the overall strategic plan at all levels.

DOES YOUR PLAN PLAN FOR THE UNEXPECTED?

Every plan developed and executed should have countermeasures identified in the event that something goes wrong during implementation. In both strategic and tactical planning, no one can foresee the future. The ability of management to identify changing variables during rollout and implementation of plans is crucial to the success of the plan. It is for this reason that many people with military backgrounds and leadership skills are highly sought for recruiting in private industry. The abilities of people with military leadership and training allow them to constantly adapt and correct plans as needed.

STRATEGIC PLANNING FOR ENVIRONMENTAL HEALTH AND SAFETY

Environmental health and safety (EHS) plans are almost always tactical in nature given the depth and breadth of some of the regulatory standards. Many companies have identified the goal of obtaining OSHA's Voluntary Protection Program (VPP) status and becoming a VPP Star facility or company. The plan to become VPP certified can be placed on the tactical plan, but many companies fail to recognize that to become a VPP site, the EHS program must be part of the strategic plan as well. To help drive strategic planning for companies that wish to go VPP, OSHA has developed the Program Evaluation Profile (PEP). The PEP analysis looks at six modules that make up an effective safety and health program. Each area is scored on a scale of 1–5, with the top score of 5 representing a world-class safety and health program.

1. No program or ineffective program
2. Developmental program
3. Basic program
4. Superior program
5. Outstanding program

The PEP analysis looks at the following modules:

- Management leadership and employee participation
  - Management leadership
  - Employee participation
  - Implementation (tools provided by management, including budget, information, personnel, assigned responsibility, adequate expertise and authority, line accountability, and program review procedures)
  - Contractor safety
- Workplace analysis
  - Survey and hazard analysis
  - Inspection
  - Reporting
- Accident and record analysis
  - Investigation of accidents and near-miss incidents
  - Data analysis
- Hazard prevention and control
  - Hazard control
  - Maintenance
  - Medical program

- Emergency response
  - Emergency preparedness
  - First aid
- Safety and health training (as a whole)

The expectation of a safety and health program from OSHA is that safety and health is permeated throughout an organization or location. It is for this reason that the VPP program is difficult to obtain. Safety and health is ingrained in the culture at VPP sites, with commitment from management as well as employee engagement.

SPECIFIC QUESTIONS AND AREAS UNDER THE PEP EVALUATION

1. Management Leadership and Employee Participation

1.1. Visible management leadership provides the motivating force for an effective safety and health program.

1. Management demonstrates no policy, goals, objectives, or interest in safety and health issues at this worksite.
2. Management sets and communicates safety and health policy and goals, but remains detached from all other safety and health efforts.
3. Management follows all safety and health rules, and gives visible support to the safety and health efforts of others.
4. Management participates in significant aspects of the site's safety and health program, such as site inspections, incident reviews, and program reviews. Incentive programs that discourage reporting of accidents, symptoms, injuries, or hazards are absent. Other incentive programs may be present.
5. Site safety and health issues are regularly included on agendas of management operations meetings. Management clearly demonstrates—by involvement, support, and example—the primary importance of safety and health for everyone on the worksite. Performance is consistent and sustained or has improved over time.

1.2. Employee participation provides a means through which workers identify hazards, recommend and monitor abatement, and otherwise participate in their own protection.

1. Worker participation in workplace safety and health concerns is not encouraged. Incentive programs are present that have the effect of discouraging reporting of incidents, injuries, potential hazards, or symptoms. Employees/employee representatives are not involved in the safety and health program.
2. Workers and their representatives can participate freely in safety and health activities at the worksite without fear of reprisal. Procedures are in place for communication between employer and workers on safety and health matters. Worker rights under the Occupational Safety and Health Act to refuse or stop work that they reasonably believe involves imminent danger are understood by workers and honored by management. Workers are paid while performing safety activities.

3. Workers and their representatives are involved in the safety and health program, are involved in inspection of work area, and are permitted to observe monitoring and receive results. Workers' and representatives' right of access to information is understood by workers and recognized by management. A documented procedure is in place for raising complaints of hazards or discrimination and receiving timely employer responses.

4. Workers and their representatives participate in workplace analysis, inspections and investigations, and development of control strategies throughout facility, and have necessary training and education to participate in such activities. Workers and their representatives have access to all pertinent health and safety information, including safety reports and audits. Workers are informed of their right to refuse job assignments that pose serious hazards to themselves pending management response.

5. Workers and their representatives participate fully in development of the safety and health program and conduct of training and education. Workers participate in audits, program reviews conducted by management or third parties, and collection of samples for monitoring purposes, and have necessary training and education to participate in such activities. Employer encourages and authorizes employees to stop activities that present potentially serious safety and health hazards.

1.3. Implementation means tools, provided by management, which includes budget, information, personnel, assigned responsibility, adequate expertise and authority, means to hold responsible persons accountable (line accountability), and program review procedures.

1. Tools to implement a safety and health program are inadequate or missing.

2. Some tools to implement a safety and health program are adequate and effectively used; others are ineffective or inadequate. Management assigns responsibility for implementing a site safety and health program to identified person(s). Management's designated representative has authority to direct abatement of hazards that can be corrected without major capital expenditure.

3. Tools to implement a safety and health program are adequate, but are not all effectively used. Management representative has some expertise in hazard recognition and applicable OSHA requirements. Management keeps or has access to applicable OSHA standards at the facility, and seeks appropriate guidance information for interpretation of OSHA standards. Management representative has authority to order/purchase safety and health equipment.

4. All tools to implement a safety and health program are more than adequate and effectively used. Written safety procedures, policies, and interpretations are updated based on reviews of the safety and health program. Safety and health expenditures,

including training costs and personnel, are identified in the facility budget. Hazard abatement is an element in management performance evaluation.

5. All tools necessary to implement a good safety and health program are more than adequate and effectively used. Management safety and health representative has expertise appropriate to facility size and process, and has access to professional advice when needed. Safety and health budgets and funding procedures are reviewed periodically for adequacy.

1.4. Contractor safety requires an effective safety and health program that protects all personnel on the worksite, including the employees of contractors and subcontractors. It is the responsibility of management to address contractor safety.

1. Management makes no provision to include contractors within the scope of the worksite's safety and health program.

2. Management policy requires contractor to conform to OSHA regulations and other legal requirements.

3. Management designates a representative to monitor contractor safety and health practices, and that individual has authority to stop contractor practices that expose host or contractor employees to hazards. Management informs contractor and employees of hazards present at the facility.

4. Management investigates a contractor's safety and health record as one of the bidding criteria.

5. The site's safety and health program ensures protection of everyone employed at the worksite, i.e., regular full-time employees, contractors, and temporary and part-time employees.

2. Survey and Hazard Analysis

2.1. An effective, proactive safety and health program will seek to identify and analyze all hazards. In large or complex workplaces, components of such analysis are the comprehensive survey and analyses of job hazards and changes in conditions.

1. No system or requirement exists for hazard review of planned/changed/new operations. There is no evidence of a comprehensive survey for safety or health hazards or for routine job hazard analysis.

2. Surveys for violations of standards are conducted by knowledgeable person(s), but only in response to accidents or complaints. The employer has identified principal OSHA standards that apply to the worksite.

3. Process, task, and environmental surveys are conducted by knowledgeable person(s) and updated as needed and as required by applicable standards. Current hazard analyses are written (where appropriate) for all high-hazard jobs and processes; analyses are communicated to and understood by affected employees. Hazard analyses are conducted for jobs/ tasks/workstations where injury or illnesses have been recorded.

4. Methodical surveys are conducted periodically and drive appropriate corrective action. Initial surveys are conducted by a qualified professional. Current hazard analyses are documented for all work areas and are communicated and available to all the workforce; knowledgeable persons review all planned/changed/new facilities, processes, materials, or equipment.

5. Regular surveys including documented comprehensive workplace hazard evaluations are conducted by certified safety and health professional or professional engineer, etc. Corrective action is documented and hazard inventories are updated. Hazard analysis is integrated into the design, development, implementation, and changing of all processes and work practices.

2.2. Inspection: To identify new or previously missed hazards and failures in hazard controls, an effective safety and health program will include regular site inspections.

1. No routine physical inspection of the workplace and equipment is conducted.

2. Supervisors dedicate time to observing work practices and other safety and health conditions in work areas where they have responsibility.

3. Competent personnel conduct inspections with appropriate involvement of employees. Items in need of correction are documented. Inspections include compliance with relevant OSHA standards. Time periods for correction are set.

4. Inspections are conducted by specifically trained employees, and all items are corrected promptly and appropriately. Workplace inspections are planned, with key observations or checkpoints defined and results documented. Persons conducting inspections have specific training in hazard identification applicable to the facility. Corrections are documented through follow-up inspections. Results are available to workers.

5. Inspections are planned and overseen by certified safety or health professionals. Statistically valid random audits of compliance with all elements of the safety and health program are conducted. Observations are analyzed to evaluate progress.

2.3. A reliable hazard reporting system enables employees, without fear of reprisal, to notify management of conditions that appear hazardous and to receive timely and appropriate responses.

1. No formal hazard reporting system exists, or employees are reluctant to report hazards.

2. Employees are instructed to report hazards to management. Supervisors are instructed and are aware of a procedure for evaluating and responding to such reports. Employees use the system with no risk of reprisals.

3. A formal system for hazard reporting exists. Employee reports of hazards are documented, corrective action is scheduled, and records are maintained.

4. Employees are periodically instructed in hazard identification and reporting procedures. Management conducts surveys of employee observations of hazards to ensure that the system is working. Results are documented.

5. Management responds to reports of hazards in writing within specified time frames. The workforce readily identifies and self-corrects hazards; they are supported by management when they do so.

3. Accident Investigation

3.1. An effective program will provide for investigation of accidents and near miss incidents, so that their causes, and the means for their prevention, are identified.

1. No investigation of accidents, injuries, near misses, or other incidents is conducted.

2. Some investigation of incidents takes place, but root cause may not be identified, and correction may be inconsistent. Supervisors prepare injury reports for lost time cases.

3. OSHA 301 is completed for all recordable incidents. Reports are generally prepared with cause identification and corrective measures prescribed.

4. OSHA-recordable incidents are always investigated, and effective prevention is implemented. Reports and recommendations are available to employees. Quality and completeness of investigations are systematically reviewed by trained safety personnel.

5. All loss-producing accidents and near misses are investigated for root causes by teams or individuals that include trained safety personnel and employees.

3.2. Data analysis: An effective program will analyze injury and illness records for indications of sources and locations of hazards, and jobs that experience higher numbers of injuries. By analyzing injury and illness trends over time, patterns with common causes can be identified and prevented.

1. Little or no analysis of injury/illness records; records (OSHA 300/301, exposure monitoring) are kept or conducted.

2. Data is collected and analyzed, but not widely used for prevention. OSHA 301 is completed for all recordable cases. Exposure records and analyses are organized and are available to safety personnel.

3. Injury/illness logs and exposure records are kept correctly, are audited by facility personnel, and are essentially accurate and complete. Rates are calculated so as to identify high-risk areas and jobs. Workers' compensation claim records are analyzed and the results used in the program. Significant analytical findings are used for prevention.

4. Employer can identify the frequent and most severe problem areas, the high-risk areas and job classifications, and any exposures responsible for OSHA-recordable cases. Data are fully analyzed and effectively communicated to employees. Illness/injury data are audited and certified by a responsible person.

5. All levels of management and the workforce are aware of results of data analyses and resulting preventive activity. External audits of accuracy of injury and illness data, including review of all available data sources, are conducted. Scientific analysis of health information, including non-occupational databases, is included where appropriate in the program.

4. Hazard Prevention and Control

4.1. Hazard Control: Workforce exposure to all current and potential hazards should be prevented or controlled by using engineering controls wherever feasible and appropriate, work practices and administrative controls, and personal protective equipment (PPE).

1. Hazard control is seriously lacking or absent from the facility.

2. Hazard controls are generally in place, but effectiveness and completeness vary. Serious hazards may still exist. Employer has achieved general compliance with applicable OSHA standards regarding hazards with a significant probability of causing serious physical harm. Hazards that have caused past injuries in the facility have been corrected.

3. Appropriate controls (engineering, work practice, and administrative controls, and PPE) are in place for significant hazards. Some serious hazards may exist. Employer is generally in compliance with voluntary standards, industry practices, and manufacturers' and suppliers' safety recommendations. Documented reviews of needs for machine guarding, energy lockout, ergonomics, materials handling, bloodborne pathogens, confined space, hazard communication, and other generally applicable standards have been conducted. The overall program tolerates occasional deviations.

4. Hazard controls are fully in place, and are known and supported by the workforce. Few serious hazards exist. The employer requires strict and complete compliance with all OSHA, consensus, and industry standards and recommendations. All deviations are identified and causes determined.

5. Hazard controls are fully in place and continually improved upon based on workplace experience and general knowledge. Documented reviews of needs are conducted by certified health and safety professionals or professional engineers, etc.

4.2. Maintenance: An effective safety and health program will provide for facility and equipment maintenance, so that hazardous breakdowns are prevented.

1. No preventive maintenance program is in place; breakdown maintenance is the rule.
2. There is a preventive maintenance schedule, but it does not cover everything and may be allowed to slide or performance is not documented. Safety devices on machinery and equipment are generally checked before each production shift.
3. A preventive maintenance schedule is implemented for areas where it is most needed; it is followed under normal circumstances. Manufacturers' and industry recommendations and consensus standards for maintenance frequency are complied with. Breakdown repairs for safety-related items are expedited. Safety device checks are documented. Ventilation system function is observed periodically.
4. The employer has effectively implemented a preventive maintenance schedule that applies to all equipment. Facility experience is used to improve safety-related preventative maintenance scheduling.
5. There is a comprehensive safety and preventive maintenance program that maximizes equipment reliability.

4.3. Medical Program: An effective safety and health program will include a suitable medical program where it is appropriate for the size and nature of the workplace and its hazards.
1. Employer is unaware of, or unresponsive to, medical needs. Required medical surveillance, monitoring, and reporting are absent or inadequate.
2. Required medical surveillance, monitoring, removal, and reporting responsibilities for applicable standards are assigned and carried out, but results may be incomplete or inadequate.
3. Medical surveillance, removal, monitoring, and reporting comply with applicable standards. Employees report early signs/symptoms of job-related injury or illness and receive appropriate treatment.
4. Health care providers provide follow-up on employee treatment protocols and are involved in hazard identification and control in the workplace. Medical surveillance addresses conditions not covered by specific standards. Employee concerns about medical treatment are documented and responded to.
5. Health care providers are on-site for all production shifts and are involved in hazard identification and training. Health care providers periodically observe the work areas and activities and are fully involved in hazard identification and training.

5. Emergency Preparedness

5.1. Emergency preparedness: There should be appropriate planning, training/drills, and equipment for response to emergencies. *Note*: In some facilities the employer plan is to evacuate and call the fire department. In such cases, only applicable items listed below should be considered.

1. Little or no effective effort to prepare for emergencies.
2. Emergency response plans for fire, chemical, and weather emergencies as required by 29 CFR 1910.38, 1910.120, or 1926.35 are present. Training is conducted as required by the applicable standard. Some deficiencies may exist.
3. Emergency response plans have been prepared by persons with specific training. Appropriate alarm systems are present. Employees are trained in emergency procedures. The emergency response extends to spills and incidents in routine production. Adequate supply of spill control and PPE appropriate to hazards on-site is available.
4. Evacuation drills are conducted no less than annually. The plan is reviewed by a qualified safety and health professional.
5. Designated emergency response team with adequate training is on-site. All potential emergencies have been identified. Plan is reviewed by the local fire department. Plan and performance are reevaluated at least annually and after each significant incident. Procedures for terminating an emergency response condition are clearly defined.

5.2. First aid/emergency care should be readily available to minimize harm if an injury or illness occurs.
1. Neither on-site nor nearby community aid (e.g., emergency room) can be ensured.
2. Either on-site or nearby community aid is available on every shift.
3. Personnel with appropriate first aid skills commensurate with likely hazards in the workplace and as required by OSHA standards (e.g., 1910.151, 1926.23) are available. Management documents and evaluates response time on a continuing basis.
4. Personnel with *certified* first aid skills are always available on-site; their level of training is appropriate to the hazards of the work being done. Adequacy of first aid is formally reviewed after significant incidents.
5. Personnel trained in advanced first aid and/or emergency medical care are always available on-site. In larger facilities a health care provider is on-site for each production shift.

6. Safety and Health Training

6.1. Safety and health training should cover the safety and health responsibilities of all personnel who work at the site or affect its operations. It is most effective when incorporated into other training about performance requirements and job practices. It should include all subjects and areas necessary to address the hazards at the site.
1. Facility depends on experience and peer training to meet needs.
2. Managers/supervisors demonstrate little or no involvement in safety and health training responsibilities. Some orientation training is given to new hires. Some safety training materials (e.g., pamphlets,

posters, videotapes) are available or are used periodically at safety meetings, but there is little or no documentation of training or assessment of worker knowledge in this area. Managers generally demonstrate awareness of safety and health responsibilities, but have limited training themselves or involvement in the site's training program.

3. Training includes OSHA rights and access to information. Training required by applicable standards is provided to all site employees. Supervisors and managers attend training in all subjects provided to employees under their direction. Employees can generally demonstrate the skills/knowledge necessary to perform their jobs safely. Records of training are kept and training is evaluated to ensure that it is effective.

4. Knowledgeable persons conduct safety and health training that is scheduled, assessed, and documented, and addresses all necessary technical topics. Employees are trained to recognize hazards, violations of OSHA standards, and facility practices. Employees are trained to report violations to management. All site employees—including supervisors and managers—can generally demonstrate preparedness for participation in the overall safety and health program. There are easily retrievable scheduling and record keeping systems.

5. Knowledgeable persons conduct safety and health training that is scheduled, assessed, and documented. Training covers all necessary topics and situations, and includes all persons working at the site (hourly employees, supervisors, managers, contractors, part-time and temporary employees). Employees participate in creating site-specific training methods and materials. Employees are trained to recognize inadequate responses to reported program violations. Retrievable record-keeping system provides for appropriate retraining, makeup training, and modifications to training as the result of evaluations.

The goal of PEP is for organizations to identify and understand what a total strategic plan for occupational safety and health is. While not everyone agrees with the Voluntary Protection Program, this tool is a good way to identify gaps in the safety culture of an organization.

SWOT ANALYSIS

SWOT analysis is another technique or tool used by organizations and companies during strategic planning. SWOT stands for *strengths, weaknesses, opportunities,* and *threats.* To help in the decision-making process during strategic planning, each of these areas should be reviewed to identify the needs of the organization. Strengths and weaknesses are usually identified as internal factors that the organization has direct control over. When strengths and weaknesses have been identified, decisions

TABLE 2.1
Example of a SWOT Analysis Table

	Helpful to Achieving Goal	Harmful to Achieving Goal
Internal	Strengths	Weaknesses
External	Opportunities	Threats

can be made regarding continuous improvement. What does the company do well or what are its strengths? Where does the company struggle, or what are its weaknesses? Companies that are engaged in lean enterprise may identify similar processes internally with very different results. Internal benchmarking and communication will help eliminate internal weaknesses in a process.

Opportunities and threats are normally considered external factors that the company or organization does not have control over. Opportunities can include new product placement, increasing market share of current business, and acquisitions of competitors. Threats can include any threat to a business model including business interruption, social unrest, and product recall. A simple box model of a SWOT analysis can help everyone understand how to identify issues in each area. SWOT analysis must start with a vision of a future or ideal state to judge all strengths, weaknesses, opportunities, and threats. A simple table can be set up to determine a SWOT, as shown in Table 2.1.

FOUNDATIONS OF DECISION MAKING

Over the years of changing management styles and paradigm shifts, problem-solving models have changed as well. This change has occurred due to continuous improvement of different processes and the skill sets of employees. Many companies will not solely rely on one method of problem solving, but rather choose different methods to solve different problems. Many people within the lean enterprise community have identified Dr. W. Edwards Deming as the grandfather of problem solving.

Deming made a process that was first used by Sir Francis Bacon in the late 1500s. This process was improved upon by Walter Shewhart and made popular by Deming. The Deming cycle of problem solving was termed PDCA (plan, do, check, act). (See Figure 2.1.)

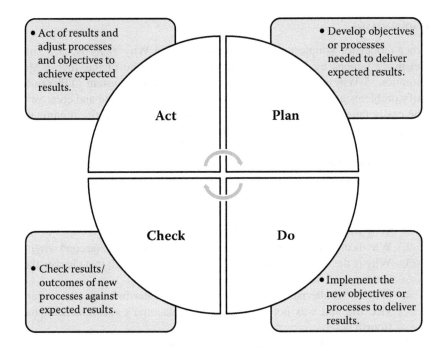

FIGURE 2.1 PDCA Cycle

Plan—Establish the objectives and processes necessary to deliver results to an expected outcome.

Do—Implement the new processes. Small-scale tests are preferred, to test possible effects of a new process.

Check—Measure all new processes, comparing the results against the expected results.

Act—Analyze the differences to determine root cause as to why the expected result was not achieved. Changes should be applied to improve the process.

If the problem statement has passed through these four steps and an improvement has not been identified, the scope of the problem and/or problem statement should be reviewed. One or more issues have not been properly identified in the problem statement. PDCA can be applied until there is a plan that involves improvement. When kaizen events are held for different process improvements, no matter how long or involved the kaizen event is, it all comes back to the PDCA. How fast can PDCA be implemented for a problem in a process? As companies move through continuous improvement, many processes will come full circle, meaning that eventually processes that have already had kaizen events for improvement will be reviewed again. The PDCA cycle will be implemented once again to identify corrective action and continuous improvement. As the PDCA cycle is repeated, more improvement should be seen as different processes are implemented.

5-Why Analysis

Another model for simple problem solving, called 5-Why analysis, is a simple model made famous by Sakichi Toyoda and Taiichi Ohno of the Toyota Motor Corporation. 5-Why was added to the Toyota Production System (TPS) to help identify problems and corrective action in a process. This simple and creative tool can be easily taught to everyone on the shop floor and in the entire organization. Starting the process is as simple as picking a problem and asking, "Why is it so?" in five successive iterations. If the 5-Why is performed correctly, the proper root cause will be determined.

The following example demonstrates the basic process:

I went to start my car this morning and it will not start. (Problem Statement)

1. Why won't the car start? The battery is dead. (first why)
2. Why is my battery dead? The alternator is not working. (second why)
3. Why is the alternator not working? The alternator belt that charges the battery is broken. (third why)
4. Why is the alternator belt broken? The alternator belt was past its service life and was not replaced to manufacturer's recommendations. (fourth why)
5. Why wasn't the belt replaced when recommended? My car has not been maintained properly according to the service schedule. (fifth why, a root cause)

I will start maintaining my car according to the recommended service schedule. (Solution)

Although it is relatively easy to step through this process, 5-Why analysis will only drive to one root cause. When multiple root causes are to blame for a process failure or problem, 5-Why may be lacking for root cause. 5-Why questions can be added to any type of investigation into a problem to help drive to root cause findings. As stated before, 5-Why is a simple process that can be taught to everyone. 5-Why should not be used as a stand-alone problem-solving tool for all issues. Some potential failures in 5-Why analysis can include jumping to conclusions without data or without going to see where the problem is and focusing on filling out a 5-Why form instead of correcting the problem. Employees engaged in using 5-Why should always try to dig deeper when asking why questions. Not completing due diligence on the questions will produce five unrelated answers to the why questions.

A 5-Why analysis should be used for problems when the "point of cause" for the problem has been identified. Several steps should be taken before asking 5-Why questions:

Go and see the problem—Many people and organizations will jump to conclusions, asking 5-Why before understanding the problem. Not understanding the problem can lead to multiple solutions to a 1-Why question.
5W1H—Think of going back to grammar school when learning how to write. Who was involved, what happened, and why, where, when, and how did the problem occur?

- Who
  - Who saw it happen?
  - Who was involved with the issue?
- What
  - What happened?
  - What occurred?
- When
  - When did it occur?
  - When in the sequence of the process?
- Where
  - Where did it occur?
- Why
  - Why did it happen?
  - 5-Why analysis?
- How
  - How did it occur?

Observe or observation—People can tell you all day long what the problem is; until you go and observe the issue or problem you will not have a good grasp of what is occurring.

Create a fishbone to determine cause and effect—Although not always needed, creating a fishbone diagram will help document the problem from identifying causes. Once a point of cause has been determined, 5-Why analysis should identify a root cause. Many people get hung up on asking why five times when, in reality, you will most likely get to a root cause within two to four questions. A good demonstration of jumping to 5-Why analysis before understanding the point of cause can be applied to the example of the dead car battery. Without going to see the problem and observing the process of why the car wouldn't start, one might have identified the problem as low fluid in the car battery or possibly a battery drain from something left on in the car. Without the observation and understanding the process, the broken alternator belt would not have been identified.

SIX SIGMA AND THE DMAIC PROCESS

In the mid-1980s, Motorola formalized a process for continuous improvement using tools that have been identified in manufacturing and quality in the past. Six Sigma has taken on many different meanings, but the summary of all definitions would tell us that Six Sigma is considered the most accurate and effective problem-solving model. Six Sigma, as defined in the book *Six Sigma for Dummies,* is the statistical basis of the approach and methodology to address two concerns: (1) roll-up of characteristic behaviors in a process and (2) the natural increase in variation in each characteristic over the long term. Sigma refers a universal scale of how well a characteristic performs compared to what it is required to perform at. The higher the sigma number, the better the characteristic performs. At Six Sigma, the defect rate is 3.4

per million. To put this in perspective, One Sigma would yield 691,462 defects per million (Gygi, DeCarlo, and Williams 2005).

DMAIC stands for *define, measure, analyze, improve,* and *control.* Some companies have chosen to add other sections to the process, such as *R* for *replicate,* creating DMAICR. Companies that struggle with internal communication may use the replicate application if the solution can be applied to other facilities with the same process. Replicate can also be used for management of change. In replicate, questions that need to be asked include the following:

- Who else in the organization can benefit from the project findings?
- What has been done to update the corporate knowledge?
- What other processes could benefit from the project findings?
- Is the improvement sustained (six months later...)?

SIX SIGMA FOR ENVIRONMENTAL HEALTH AND SAFETY

The Six Sigma process can be applied to any problem and identify improvements needed in a process, and safety is no different. The difference in using this process for safety, health, and environmental issues is that the customer is most likely internal to your organization. If identifying a safety issue, the customer will be your employees or the company you work for. Environmental issues may trigger an outside customer, depending on what the issue is. When we use the DMAIC process for safety, we need to tweak the questions that the process poses to put a safety spin on the issue, especially when defining the problem. In reality, the DMAIC process is a pumped-up version of Deming's plan, do, check, act. Figure 2.2 shows an example of a DMAIC process for EHS.

DMAIC processes to help identify and drive safety issues can be used in many traditional and nontraditional ways. In Table 2.2, the DMAIC process is used as a quick review of safety and health issues to make sure that department managers at one location know and understand safety issues that occurred in their respective departments.

6M problem solving is a model used to identify cause and effect for a given problem and is considered a typical problem-solving tool. The six M's refer to each area that should be addressed as a cause of the problem. The M's stand for the following:

1. Man—human variation: person to person, shift to shift (also could be *mammal* instead of *man* to be politically correct)
2. Machine—machine warm-up, shutdown, settings, changeover
3. Material—age of material, dimensions, specifications
4. Method—standard work, work instructions
5. Mother Nature—environmental factors, humidity, etc.
6. Measurement system—are we measuring it correctly?

Each cause is identified in a fishbone diagram falling under the heading of one of the six M's, as shown in this diagram. Other categories have been identified within the cause-and-effect problem-solving process. These include categories for developing

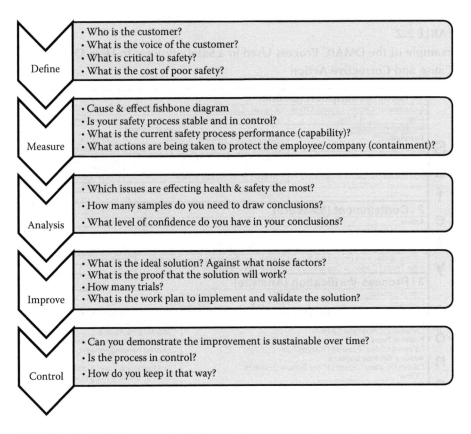

FIGURE 2.2 DMAIC process for EHS example.

fishbone diagrams recommended for administration and service industries using an 8P system that includes price, promotion, people, processes, place, policies, and procedures/product. Figure 2.3 demonstrates how the fishbone diagram is set up for problem solving.

PDCA has been identified as the most basic problem-solving and improvement tool by many. Companies have incorporated the PDCA cycle into other, more complex problem-solving models such as the Eight Disciplines of Problem Solving or 8D. 8D problem solving was developed in the mid-1980s by Ford Motor Company's Power Train Operations (PTO) division (Wikipedia 2011).

1. Use a team: Establish a team of people with product/process knowledge.
2. Define and describe the problem: Specify the problem by identifying in quantifiable terms the who, what, where, when, why, how, and how many (5W2H) for the problem.
3. Develop an interim containment plan; implement and verify interim actions: Define and implement containment actions to isolate the problem from any customer.

TABLE 2.2

Example of the DMAIC Process Used in a Safety Process to Help Drive Root Cause and Corrective Action

Safety Concern Sheet

1 Issue Description (Define)

Department: ☐Body ☐Paint ☐Trim ☐Chassis ☐Quality ☐MP&L ☐Skilled Trades			**Shift:** ☐1 ☐2 ☐3
Date:	Time:	Time Employee Began Work:	Bay Location:
Workstation Location:		**Type of Contact**	
Description:		☐Caught in, on, between, or under	☐Contact with Electricity
		☐Exposed to harmful conditions or substance	☐Exposure to Noise
		☐Exposure to extreme temperatures	☐Fall or jump to below
		☐Overexertion-Acute	☐Overexertion-Repetitive
		☐Rubbed or abraded by friction	☐Slip/Trip/Fall
		☐Struck Against	☐Struck by
		☐Other:	

2 Containment (Measure)

Action:		
Containment Currently in Place? ☐Yes ☐No	Workstation Location:	Bay Location:
Will the containment prevent the Type of Contact identified in the Issue Description? ☐Yes ☐No		
If the Type of Contact is Overexertion-Repetitive, has Ergonomics been notified? ☐Yes ☐No		

3 Process Verification (Analyze)

Area Supervisor:	WGL:	Operator:
Operation:	Process #:	Bay Location:

Injury Source	Task/Activity
Material Handling	☐Maintenance/Repair-Breakdown
☐Manual ☐Crane/Hoist ☐PMHV	☐Maintenance Routine
Portable Tools	☐Manual Assembly or Disassembly
☐Powered ☐Non-Powered ☐Cutting Tool	☐Material Handling, including PMHV
Walking Working Surfaces	☐Office Tasks
☐Stairs ☐Ladder ☐Ramp ☐Floor Surface ☐Platform	☐Not Performing Tasks (walking, bathroom)
☐Other:	☐Driving, operating riding in vehicle
Manual Assembling/Disassembling Parts	☐Operating Machine/Tooling/Equipment
☐Fasteners ☐Connectors ☐Clamps ☐Bolts ☐Screws	☐Unknown
☐Other:	☐Other:

JSA	Reserved
PPE correct for the task? ☐Yes ☐No	
Is the operator properly training for the task? ☐Yes ☐No	
Are any/all hazardous chemicals notes on JSA? ☐Yes ☐No ☐NA	

4 Permanent Corrective Actions (Improve)

Root Cause Understood? ☐Yes ☐No		**Corrective Action**
Personal Factors	**Job Factors**	☐Education
☐Inappropriate Work Assignment	☐Leadership:	☐Enforcement
☐Lack of Appropriate Training	☐Problems in Facility Design, Engineering	☐Engineering
☐Stress	☐Maintenance Wear & Tear	☐Maintenance
☐Motivation	☐Problems with Tools & Equipment	☐Counseling/Advisement
☐Abuse/Misuse of Tools & Equipment	☐Problems with Standards or Procedures	☐Other:

Incident Root Cause	Interim Corrective Action	Permanent Corrective Action
	Target Date:	Target Date:

Has the corrective action been communicated to other shifts? ☐Yes ☐No	Who was contacted:
Superintendent Review:	**Date Reviewed**
Department Manager Review:	
Safety Review:	

4. Determine, identify, and verify root causes and escape points: Identify all potential causes that could explain why the problem occurred. Also identify why the problem was not noticed at the time it occurred. All causes shall be verified or proved, not determined by fuzzy brainstorming.

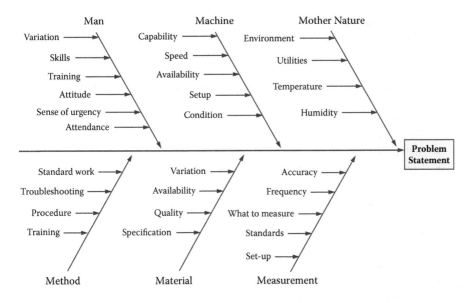

FIGURE 2.3   Cause-and-effect diagram, commonly known as a fishbone diagram.

5. Choose and verify permanent corrective actions (PCAs) for root cause and escape point; through preproduction programs quantitatively confirm that the selected corrective actions will resolve the problem for the customer.
6. Implement and validate PCAs: Define and implement the best corrective actions.
7. Prevent recurrence: Modify the management systems, operation systems, practices, and procedures to prevent recurrence of this and all similar problems.
8. Congratulate your team: Recognize the collective efforts of the team. The team needs to be formally thanked by the organization.

8D has become a standard in the auto, assembly, and other industries that require a thorough structured problem-solving process using a team approach. The 8D process takes several problem-solving tools and combines them into one uniform model. It is for this reason that many companies do not use full 8D problem-solving techniques. The model is so comprehensive that many lean professionals steer clear of this process due to the cumbersome requirements.

KEPNER-TREGOE PROBLEM SOLVING

Dr. Charles Kepner and Dr. Benjamin Tregoe met each other while working for the RAND Corporation in 1958 as researchers. The two were asked to identify issues in the decision-making process within the Strategic Air Command (SAC) for the U.S. Air Force. As research continued into the issues, they both identified that Air Force personnel were not gathering, organizing, and analyzing data properly before making decisions and taking corrective action. Kepner-Tregoe (KT) process improvements include rational thinking modules pushing that managers in business must understand what is happening in the real world of their process. This also ties into lean enterprise, as companies must be brutally honest with themselves to truly understand the current state of the business and market. KT teaches the following:

- Defining the process (how it is done)
- Measuring process performance (establishing a baseline)
- Stabilizing the process (removing variation)
- Improving the process (how it should be done)

KT processes include the following:

- Situation appraisal—clarifies the current status
- Problem analysis—identifies the root cause of process breakdowns and deviations
- Potential problem/opportunity analysis—ensures that implementation and continuous improvement efforts are successful
- Project management—guides the project from definition through implementation

KT decision-making processes and models are used and have been used at Fortune 100 companies throughout the world. These companies include the Honda Motor Company, BASF (AG), Motorola, and ExxonMobil Corporation (Kepner-Tregoe, Inc. 2010).

ROOT CAUSE ANALYSIS AND INCIDENT INVESTIGATION— CASE STUDY OF FORD MOTOR COMPANY

During my time at Ford Motor Company (English 2006), I worked on a project to identify potential issues with the current incident investigation modules that Ford was using. I wrote a piece for *Occupational Hazards* that appeared in the December 2006 issue titled "Ford Motor Co. Rolls Out a New Safety Model." That article was produced from the following full account of the research and findings for this project.

Using 5-Why analysis for incident investigations involving safety will help supervisors and managers drive to root cause and can be used as learning and coaching tools for lean enterprise and safety. Many times during injury investigations, many people misdiagnose root cause and implement corrective action that is not needed. Many large companies have some type of root cause analysis built into the incident investigation process to help determine root cause. What has been found in recent years is that even with a root cause model built into the process, management investigating injuries still must understand how to drive for root cause. A study completed at Ford Motor Company in 2006 reviewed incident investigations at different facilities and found that first-time quality of incident investigations was at best lacking for root cause analysis.

In 2000, Ford Motor Company implemented the Occupational Health and Safety Information Management System (OHSIM). This was established to provide a uniform, consistent tool to provide the company with accurate incident investigations regarding workplace illnesses and injuries. Facilities were using different media and methods to collect workplace illness and injury data up until OHSIM was implemented. Before OHSIM, the process of incident investigation met the needs of the record-keeping standard for OSHA but did not address the needs of business to identify root cause and help prevent injuries. It was determined that the use of one corporate-wide database would benefit the company to help understand all illnesses and injuries and drive for root cause.

At one Ford location, the plant attempted to fully utilize the OHSIM process to identify injuries and illnesses in the workplace by identifying workstations and job tasks with a high frequency of incidents. The facility reviewed current data regarding job process codes within the OHSIM system. Job process codes are alphanumeric codes that are assigned to each workstation to identify a task or job. Job process codes were reviewed for any type of trends for injuries and/or illnesses. These injuries include first aid injury data, which comprise first-time office visits (FTOVs). FTOVs occur when an employee requires medical attention for a workplace injury, illness, or discomfort.

Upon review it was determined that 20% of all FTOVs occurring between January and September of 2005 were unidentifiable due to the absence of a job process code. Management determined that to properly identify unsafe acts and conditions, a quality control process needed to be established regarding OHSIM investigations by line supervisors. Safety identified six top issues regarding the quality of the OHSIM incident investigations. These issues were set up as rejection criteria during the review or safety sign-off phase of the incident investigation. Proper corrective action in response to the incident would ensure that the supervisor would disposition the incident correctly, and the unsafe act or condition would be corrected.

Overall quality was the first rejection criterion to be utilized. Investigations would be rejected if the investigation was overall poorly written. This included, but was not limited to, misspelled words, vague investigator statements, and personal opinion written into the investigation statement.

The second rejection criterion to be utilized is the absence of a correct job process code or location code for the incident. Each production workstation is assigned an alphanumerical code that can be entered into the OHSIM investigation. The job process code ensures that every workplace injury and illness recorded can be linked to a workstation and/or task. Service departments such as quality, maintenance, and human resources are not limited to a job process code and move freely throughout the facility. If a service department employee becomes injured, a location code is required in place of the job process code. The location code will determine where the injury occurred, identifying trends for high-frequency areas within the facility.

The third rejection criterion used in the quality control process is changes to the operator instruction sheet (OIS) and job safety analysis (JSA). Each workstation has a written OIS/JSA to identify how to complete the job task and the hazards associated with each job task step. If the corrective action involved a process change, the change must be noted and changed on the OIS/JSA package at the workstation. If the investigation noted a change to the process, including personal protective equipment

(PPE), changes to the OIS/JSA must be noted and documented in the corrective action section of the OHSIM investigation.

The fourth rejection criterion to be reviewed is the abuse of coaching or counseling an employee for an unsafe act or condition. If the Safety Department identifies a physical adjustment to the work environment, the investigation would be rejected with a recommendation for the adjustment. If the investigation identifies a violation of a safety policy, process, or procedure, the investigation would be rejected with a recommendation for enforcement of controlling safety programs.

The fifth rejection criterion regarded the type of contact the investigator chose as part of the incident analysis. If the investigator identifies the wrong type of contact for the incident, the investigation would be rejected with recommendations on what the type of contact should be changed to. The following list shows the "Type of Contact" section within the OHSIM incident investigation process. The investigator can choose from 12 predetermined definitions for the type of contact identified for the incident, as well as an "Other" category to capture more variable data as to the type of contact sustained by the injured employee.

- Caught in, on, between, or under
- Contact with electricity
- Exposure to harmful conditions or substances
- Exposure to noise
- Exposure to temperature extremes (burns/frostbite)
- Fall or jump to below
- Overexertion—acute
- Overexertion—repetitive
- Rubbed or abraded by friction or pressure (blisters)
- Slip, trip, or fall on same level
- Struck against
- Struck by
- Other: _____

The sixth and final criterion developed to help increase the quality of the OHSIM incident investigations is the rejection of investigations with no action on recommendations. This means that the investigation was previously rejected for another criterion mentioned previously in this report. These investigations are rejected because the supervisor did not follow or implement the recommendations of the supervisor's superintendent and/or the Safety Department.

Once the rejection criterion was determined, incident investigations were reviewed by the Safety Department using the new rejection process. It was determined that the facility would be audited for first-time quality of the OHSIM incident investigations. Investigations would only be rejected once, unless recommendations made by management were not followed. With the new quality control process in place, the facility was shown to have a first-time quality pass rate of 63%.

The overwhelming reason why investigations were being rejected by the Safety Department was the overall quality of the report. Forty-six percent of all investigations were written poorly, 18% had no process code or work location assigned, and 12% had coaching and counseling employees as corrective action when unsafe conditions were identified but not corrected. The initial intent of the quality control process was to identify and ensure proper job process codes and location codes for each investigation. During the review of this first report in the process, it was clearly evident that training on proper, clear, and concise incident investigation was needed for employees with direct reports who are responsible for conducting incident investigations.

A single point lesson (SPL) was developed and distributed to all employees regarding the quality of the incident investigations. The SPL reviewed bullet points that can help an investigator improve the quality of the OHSIM investigation to prevent rejection of the investigation. A meeting was then held with all department managers to determine a containment action to improve the first-time quality of the OHSIM investigations. A more precise containment was needed to determine which department and, more specifically, which supervisor was consistently having OHSIM investigations rejected. The rejection data was reevaluated to identify poor-performing departments and supervisors in those departments. During the containment meeting with the department managers, it was determined that a training program should be established to identify investigation information that was critical to investigation quality. Each department was set up for a training session regarding the reasons why OHSIM incident investigations were being rejected. Issues were identified by each department based on the same rejection criteria as used for the entire facility.

Reviewing the information in specific cases in a department having the worst quality for incident investigations revealed that investigations were being rejected for the overall poor quality of the investigation itself and an abuse of coaching and counseling employees as a form of corrective action. Overall quality can be improved by paying attention to the details of the investigations. The overuse of coaching and counseling as a corrective action to an incident identified investigations in which the corrective action did not address the root cause of the incident. In many of these cases engineering, enforcement of current safety policies or procedures, and maintenance of equipment would be a more suitable corrective action.

In developing the training program to improve the first-time quality of the incident investigations, quality had to be defined as well as the customer. The following questions were identified:

- What is the definition of quality for an OHSIM incident investigation?
- Who is the customer for this quality control program?

The definition of quality for an OHSIM incident investigation must contain the following information:

- Identification of the area in which the incident occurred
- Identification of the root cause through the use of the OHSIM tools
- Containment action and/or corrective action to prevent the recurrence of the incident

The OHSIM quality control process has several different customers, each with different ideas of the definition of what quality is for OHSIM incident investigations. Each customer needed to be identified and trained on the quality control program. Three separate and distinct customers were identified during the implementation of the quality control process and subsequent training. The customers identified included management, the supervisors or investigators, and the safety department.

Customer #1, Management—Management was identified as the main customer for the quality control process. Management will drive the process to eliminate the poor quality in the OHSIM investigations. Management was the first group within the quality control process to be trained in the definitions of a quality investigation and the rejection criteria.

Customer #2, the Supervisor—The supervisors who are predominantly the investigators in the OHSIM investigation process were identified as the second customer in the quality control process. The investigators are motivated by having the incident investigation accepted and reducing the workplace injuries and illnesses of their subordinates. Supervisors only wanted to handle OHSIM investigations once, submitting a quality incident investigation. This would help supervisors manage time in a more efficient manner. During the training sessions, supervisors were encouraged to voice concerns regarding past investigations and issues surrounding the quality issues determined the quality control process. Examples of past, rejected investigations and the reasons why the investigation was rejected were reviewed. Supervisors were also required to complete the new online, Web-based OHSIM incident investigation program. The site-specific training was completed in 90 minutes. The online training program provided by Ford Motor Company Occupational Health and Safety was completed in an average of 3 hours.

Customer #3, the Safety Department—The last customer identified in the quality control process is the safety department or safety function. Safety is responsible for the final acceptance in each OHSIM investigation. Safety is motivated by identifying high hazard areas where incidents and injuries are occurring. Quality investigations will help all customers better understand and identify problem areas for safety within the facility.

Each customer had been identified and trained in the definitions of quality for an OHSIM incident investigation. Investigations were reviewed for quality each week, with the results reviewed each Friday. Each department was given several different data sets, including the following:

- Facility first-time quality results
- Department first-time quality results
- Percentage of first-time quality for department supervisors
- Reasons by percentage each department has investigations rejected

Concurrently, safety metrics were reviewed and the quality of the incident investigations improved. In the months following implantation of the OHSIM quality control process, several items were adjusted to accommodate each customer defined in the process. One department, the biggest department in the facility, was broken down into three specific and distinct departments. Each department was also reviewed on first-time quality on a weekly basis.

Significant improvement was shown in each department from the original baseline established in 2005. Improvements in first-time quality of OHSIM incident investigations included double-digit improvement in every department. A review of corrective action used since the implementation of the quality control process determined that the following occurred:

- 47% decrease in the use of coaching and counseling employees as a corrective action
- 11% decrease in the use of educating the employee in safety as a corrective action
- 123% increase in maintenance fixes and repair as a corrective action
- 10% increase in implementing engineering controls as a corrective action

Reviews of four other facilities within Ford Motor Company were sampled to determine if the quality control process would identify inferior OHSIM incident investigations for the same time period. Three of the facilities that were sampled were final assembly facilities similar to the pilot plant for the quality control process. The fourth facility was a metal stamping facility. Incident investigations were reviewed using the same quality control process as the pilot facility. First-time quality of the OHSIM incident investigations was 52.5% overall.

The same rejection criteria was applied for all OHSIM incident investigations in the sample population. Thirty-four percent of the investigations would have been rejected for ineffective changes to the operator instruction sheets (OIS) and job safety analysis (JSA). OIS/JSA packages are located at each workstation and provide any employee performing that job or task all of the instructions and noted hazards for performing that job or task. Thirty-four percent of the investigations submitted never made changes to the OIS/JSA package to identify hazards for the next shift or operator.

Recommendations made from a known and existing incident report must be communicated to all employees involved in the job task, workstation, or process. The Occupational Safety and Health Administration (OSHA) describes the use of the JSA in the corrective action phase of the incident investigation:

> Recommended preventive actions should make it very difficult, if not impossible, for the incident to recur. The investigative report should list all the ways to "foolproof" the condition or activity. Considerations of cost or engineering should not enter at this stage. The primary purpose of accident investigations is to prevent future occurrences. Beyond this immediate purpose, the information obtained through the investigation should be used to update and revise the inventory of hazards, and/or the program for hazard prevention and control. For example, the Job Safety Analysis should be revised and employees retrained to the extent that it fully reflects the recommendations made by an incident report. Implications from the root causes of the accident need to be analyzed for their impact on all other operations and procedures. (Occupational Health and Safety Administration)

Thirty-three percent of the investigations tested against the model would have been rejected for an abuse of coaching and counseling employees as a corrective action. These investigations identified corrective actions other than coaching and counseling an employee on safe behavior and work practices, but these actions were not implemented. These investigations included instances of employees failing to wear personal protective equipment but being coached and counseled on safe behavior instead of enforcing safety policy and procedure.

Seventeen percent of the investigations would have been rejected back to the supervisor for quality. Misspelled words, lack of investigation by the supervisor, and failure to identify root cause through investigation were included in this criterion. Nine percent of the investigations were rejected due to improper type of contact noted in the investigations. An example is an employee who was struck by a tool. The type of contact noted on the investigation report was a repetitive overexertion. Another 7% of the investigations would have been rejected for a substandard condition. These were investigations where the investigator identified a substandard condition during the investigation but did not identify the condition in the incident analysis.

The quality of the workplace injury investigation is a key component in the loss prevention program and model. Poor investigations will lead to the wrong conclusions regarding hazards in the workplace. In addition, recent research has determined that if a line supervisor's approach to workplace injuries was improved, it had a direct effect on reducing the severity of the injury:

> According to a new Research Institute investigation, companies that improve the way supervisors respond to employees' work-related health and safety concerns can produce significant and sustainable reductions in future injury claims and disability costs. Supervisors trained to properly respond, communicate, and problem solve with employees reduced new disability claims by 47 percent and active lost-time claims by 18 percent. The outcome measures for each group, compiled from workers compensation claims data, included the number of new and existing claims, injury types, and total indemnity costs. Both groups received similar efforts to improve workplace ergonomics. The intervention group showed a 47 percent reduction in the number of new workers' compensation claims filed after the supervisor training workshops, while the control group showed a 19 percent reduction in new claims during the same time. When the control group finally took the workshop, they experienced a further 19 percent reduction in new claims—for a total reduction of 38 percent. (Liberty Mutual 2006)

Ford Motor Company performs quarterly lean behavior surveys to check the pulse of the lean culture. Four questions concerning employee health and safety were posed to all employees. In all four questions, double-digit improvements were indicated, which represented a 25% improvement in the Ford Production System Lean Behavior Survey for hourly employees.

Questions in the lean survey were as follows:

- At this location, management is taking health and safety issues seriously.
  - 28% increase or improvement
- The health and safety training I have received helps me do my job in a healthy and safe manner.
  - 26% increase or improvement

- My supervisor consistently enforces healthy and safe work practices.
  - 21% increase or improvement
- This location is a safe and healthy place to work.
  - 23.3% increase or improvement

There is no substitute for conducting a quality incident investigation for workplace injuries and illnesses. When incident investigations are conducted properly, root cause is determined with permanent corrective action implemented to prevent further injuries. Incident investigations can identify safety trends and issues that a workstation, work zone, department, or area may be incurring. Proper incident investigation will affect the classic safety metrics of lost time, days away and restricted time (DART), and severity. Concurrently, a review of standard safety metrics demonstrated a reduction in injuries and illnesses with a 28% decrease in OSHA-recordable injuries and a 29% decrease in DART cases at the pilot facility. These metrics logically declined as hazards were identified and eliminated through the incident investigation process through determining root cause.

In making decisions and solving problems, there is no magic bullet that will work for every problem. As problems are identified and decisions are made, you must make the decision on the best method for solving the issue. In Figure 2.4, a 5-Why report is created to help identify and solve problems. The report is straightforward and easy to follow. However, questions must be asked: Who will use this report? When will it be used? What is the intent? Depending on what the expectation is for problem solving, different vehicles can be used. In Figure 2.5, a similar problem-solving form using

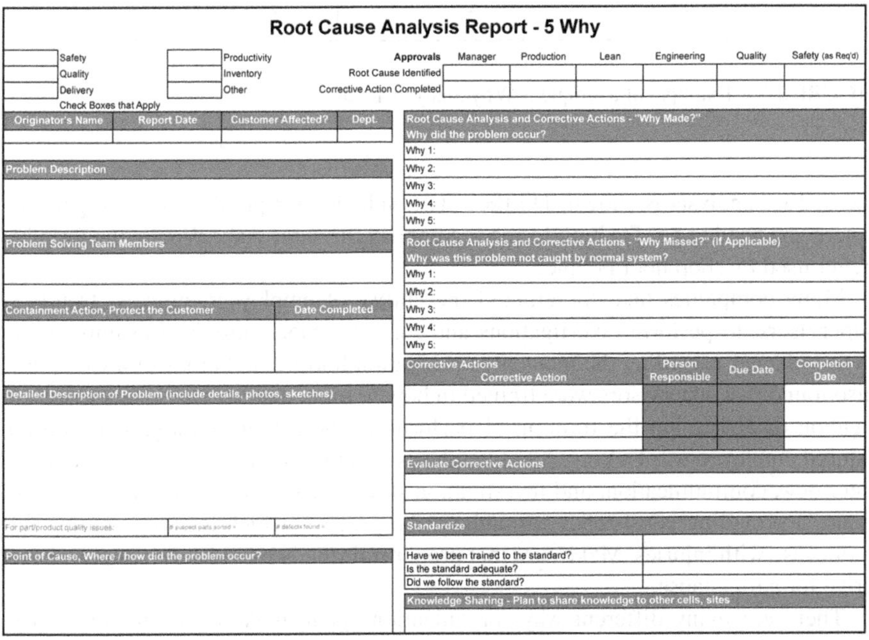

FIGURE 2.4   Root cause analysis example using 5-Why.

```
┌──────────────────────────────────────────────────────────────────────┐
│                    ROOT CAUSE ANALYSIS - 5 WHY                         │
├──────────────────────────────────────────────────────────────────────┤
│ Originator's Name:_____          Date:_____       │
│ Problem Description:                                                    │
│                                                                        │
│                                                                        │
├──────────────────────────────────────────────────────────────────────┤
│ 5 Why's - "Why Made"                                                   │
│ Why #1                                                                 │
│                                                                        │
│    Why #2                                                              │
│                                                                        │
│       Why #3                                                           │
│                                                                        │
│          Why #4                                                        │
│                                                                        │
│             Why #5                                                     │
├──────────────────────────────────────────────────────────────────────┤
│ 5 Why's - "Why Missed" (If Applicable)                                 │
│ Why #1                                                                 │
│                                                                        │
│    Why #2                                                              │
│                                                                        │
│       Why #3                                                           │
│                                                                        │
│          Why #4                                                        │
│                                                                        │
│             Why #5                                                     │
├────────────────────────────────────┬──────┬──────┬────────────────────┤
│          Corrective Actions         │ Who  │ When │       Status       │
├────────────────────────────────────┼──────┼──────┼────────────────────┤
│                                     │      │      │                    │
│                                     │      │      │                    │
│                                     │      │      │                    │
│                                     │      │      │                    │
└────────────────────────────────────┴──────┴──────┴────────────────────┘
```

FIGURE 2.5   Example of a simple 5-Why analysis form.

the 5-Why analysis is shown. Do these two vehicles for problem solving get you where you need to go? Do both forms accomplish the same task? Would one form be better used by shop floor people?

Many companies take for granted the ability of employees, namely, frontline supervisors, to perform investigations and solve for root cause. The example used in this chapter regarding the incident investigation findings at Ford Motor Company also found that supervisors were trained in how to input data into the incident investigation database, but the training of performing the actual investigation with an injured employee was lacking. More emphasis was placed on keystrokes rather than interview, communication, and investigation skills. Part of sustaining the EHS program is providing consistency throughout the process, which includes solving for root cause with injuries. Make sure that proper training is provided for employees to drive for a root cause.

There are many different ways organizations plan, make decisions, and solve problems. The job of the EHS professional is to identify what works for them and utilize the tools given by your company.

WORKS CITED

English, Paul. "Ford Motor Co. Rolls Out a New Safety Model." *EHS Today*, December 12, 2006: 48–52.

Gygi, Craig, Neil DeCarlo, and Bruce Williams. *Six Sigma for Dummies.* Indianapolis: Wiley, 2005.

Kepner-Tregoe, Inc. *Issues We Solve.* 2010. http://www.kepner-tregoe.com/resolve/Operations-BPI.cfm (accessed February 23, 2011).

Liberty Mutual. *Newsletter Volume 9, #1.* March 6, 2006. http://www.libertymutual.com (accessed December 2, 2006).

Occupational Health and Safety Administration. *Module 4 Accident/Incident Investigation eTools.* October. http://www.osha.gov (accessed October 3, 2007).

Occupational Health and Safety Administration. (1996, August 1). *Program Evaluation Profile.* (O.H. Administration, Producer) OSHA.org. http://www.osha.gov/dsg/topics/safetyhealth/pep.html (retrieved January 23, 2011).

Wikipedia. *Eight Disciplines Problem Solving.* January 25, 2011. http://en.wikipedia.org/wiki/Eight_Disciplines_Problem_Solving (accessed February 28, 2011).

3 Components of Lean Enterprise

The world we have created is a product of our thinking; it cannot be changed without changing our thinking.

Albert Einstein

Lean enterprise can consist of several different modules depending on where a particular facility or company is in its current lean journey. The idea and visual representation of the "house" comes directly from the Toyota Production System (TPS). The process of building a business or a culture is much the same as building a house. Strong foundations need to be created to build the house and continuously improve on the foundation.

When the *house of lean* is discussed, it defines what is important to building and sustaining the desired culture of the facility or company. Many companies strive to implement TPS without understanding the total philosophy of lean, spending too much time or emphasis on one function of lean or in one department. Remember, lean is a way of thinking and used to change the culture. One principle or value will not work without the other. For this reason, many people see their house of lean differently than others.

What should be included in your company's house of lean? It depends where your facility or company is in the lean journey. Lean enterprise changes as the journey progresses, just like the EHS program. If a specific target or objective has been met, does your thinking process change? What can lean do for the EHS program? The fact of the matter is that lean enterprise can help jump-start a stalled safety or environmental program and vice versa. Identifying hazards, making ergonomic improvements, and engineering hazards out of a process can all be done in the name (and budget) of lean enterprise.

VALUE STREAM MAPPING

Value stream mapping (VSM) is used to identify value-added from non-value-added work steps in a process. When value stream mapping occurs, non-value-added work should be identified and targeted for elimination or improvement as part of the lean process. Once opportunities for improvement have been identified, a kaizen will be scheduled to identify root cause and corrective action. The value stream itself is the overall big picture of the process. This map is the current state of the process. It should document what is occurring in real life and not what should be occurring. Any data flowing into a value stream that is not validated can skew the entire map. Once validated, this is what is called the *current state map*. This map shows the process, the waste, and the opportunities for improvement. Another map that should be created and reviewed is the *ideal state*. The ideal state map is a map of the process that would exist with unlimited time and resources thrown at the process. In a perfect world, what is the best application for the process? Once the current state and ideal state have been identified, the goal is to fall somewhere in the middle. Mapping out the *future state* is where the process needs to be in the near future. Depending on what the value stream map is mapping, goals and accountability can be identified at the state.

When we talk about value stream mapping, we want to map out what the customer is willing to pay for and all work that adds value to the process. Looking at the simple model in Figure 3.1 of filling a drink order in a fast-food restaurant, one can easily identify what the customer is willing to pay for and define value-added work. Figure 3.2 shows a value stream map with a large amount of data inputs. Each colored note represents a separate and distinct process.

Once a VSM is developed, areas of improvement will become self-evident.

THE KAIZEN PROCESS

More than likely you'll hear the term *kaizen* or *kaizen blitz*. What is kaizen? *Kaizen* is a Japanese term meaning "change for the better or incremental improvement." The actual act of continuous improvement can take on many different looks and approaches. When we look at identifying waste, standard work, or flow, we strive to continuously improve that process.

Different companies can have different kaizen philosophies. This is due to the fact that many companies that have started the lean journey are at different levels of lean enterprise. Many companies starting out on their lean journey will perform kaizen events at a slower pace and may have trouble identifying non-value-added work, whereas other companies have become lean savvy with self-directed workforces that need very little help performing kaizen events. The kaizen continuous improvement process is basically the plan, do, check, act (PDCA) cycle. As companies get more proficient at holding kaizen events, the faster the cycle gets. As more improvements are made to a process, the PDCA cycle gets faster.

PREPARATION FOR A KAIZEN EVENT

As stated in the beginning of the chapter, creating a VSM will ensure that all non-value-added time and steps have been identified and all issues that prevent adding value to the process have been identified. We are now ready to form a kaizen event. Key members and milestones for the project include the following:

- Event champion or sponsor—Advises team and helps remove roadblocks that might interrupt team actions
- Kaizen team leader—Leads the team, determines objectives and scope of the kaizen event, sets agenda and goals for team
- Team members—Can be internal employees or external suppliers and vendors depending on the improvement to be made

Once the scope, objective, or goals have been identified and the team has been selected, a kaizen team charter should be documented to make sure that scope creep

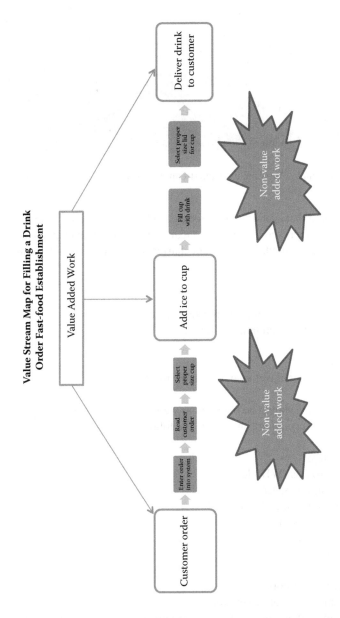

FIGURE 3.1 Example of a simple value stream map.

FIGURE 3.2   Example of a complex value stream map using a swim lane approach to mapping out multiple processes.

does not occur and that everyone stays focused. The charter should be created by the kaizen leader and approved by the kaizen team. In Figure 3.3 an example of a kaizen charter form is represented. Kaizen team charters can include the following:

- Objectives
- Scope
- Resources available (budget, assistance, etc.)
- Authority of the team (and its limits)
- Deliverables
- Schedule
- Code of conduct (developed at kick-off meeting)

Other information should be obtained as well before the team is ready to kaizen a process or area. Information can include the following:

- Process steps
- Work in process (WIP) levels
- Capacities/process times
- What is produced and how much?
- Cycle and queue times
- Batch sizes and changeover frequency
- Defect rates
- Uptime
- Number of operators
- Photos and layout of target area
- Environmental issues such as hazardous waste, process water used or discharged
- Safety issues such as past injury histories, lifting equipment, powered vehicles usage

Lean Event Project Charter

Processing Area: _____	Run Dates: _____
Champion: _____	
Event Leaders: _____	Request Date: _____

BACKGROUND (Describe The Situation)

PROBLEM STATEMENT (Define the scope, what are you trying to solve)

OBJECTIVE S.M.A.R.T. Goals (Specific, Measureable, Attainable, Realistic, and Time Based)

IMPROVEMENT METRICS

	Item	Current	Target	% Change
1				
2				
3				
4				

TEAM MEMBERS and DEPARTMENTS

	Name	Department		Name	Department
1			8		
2			9		
3			10		
4			11		
5			12		
6			13		
7			14		

POTENTIAL ROADBLOCKS

LEAN EVENT APPROVALS

| Vice President of Area: _____ Date | Director of Area: _____ Date |
| Director of Lean: _____ Date | Superintendant / Manager of Area: _____ Date |

Routing: Superintendant --> Director --> VP --> Lean --> Post in Area Impacted

FIGURE 3.3 Example of a kaizen project charter form.

Any time a kaizen occurs within your organization, a safety and health professional needs to get involved. There is a very good chance that when continuous improvement is made, there is an impact on employee safety and health that is either positive or negative. Most likely, no one in your organization sets out to impact safety and health negatively. There is a chance that during a kaizen, what is viewed as a hazard can be eliminated only to create a different or more dangerous hazard. When a kaizen is performed, safety and health should be reviewed during every aspect of the event.

Depending on how the safety and health department is structured, the chances of making all kaizen events are slim. The ability to train other employees in your organization in good EHS work practices will help foster lean change throughout other kaizen events. Although many companies train employees on the proper safety standards and procedures, the retention of that information may not transfer to the shop floor. EHS will need to identify frequent and redundant safety issues that may arise during the kaizen process. When many companies begin the lean journey, common safety questions will occur regarding everything from hazardous materials to color-coding pipes. Developing standard safety work for the kaizen team is just as important as developing standard work for the shop employee.

WASTE ELIMINATION

Waste can be found in almost every area of a process if you know what to look for and what you are looking at. Many people in lean have identified seven basic different kinds of waste that are most common in companies. Without breaking down each type of waste, different people will have different observations on what waste is. Waste can include the following:

- Too many process steps
- Excessive travel distance or time
- Waiting
- Ineffective scheduling
- Excessive handling
- Inventory storage areas
- Excess WIP
- Bottlenecks
- Defects
- Poor organization of work area (5S needed)
- Large batch sizes and long setup times
- Disjointed process steps (need flow)
- Inefficient processes
- Opportunities to apply technology to improve efficiency

The goal of this book is to help identify how safety and environmental issues can be identified and eliminated. Harnessing the power of lean will help reduce injuries, illnesses, and environmental footprints. EHS programs will continuously improve along with other business support groups during lean initiatives.

CORRECTION WASTE

Correction of a product or service due to rework, defects, or scrap is obviously considered a form of waste. However you want to categorize it, there is no substitution for first-time quality. Any time something needs to be corrected it is considered non-value-added work. The customer doesn't care and doesn't want to pay for product or service failures. What does the customer require to be fulfilled? If you can answer this question, chances are you're a genius. Before working in the automotive sector, I never realized what a gamble launching a new vehicle could be until you are in the middle of it. In most cases, to design a car from concept to a "designed for manu-facturability" platform takes 2–3 years. Can you predict what a customer will want in 2–3 years? Purchasing a vehicle is usually the second-largest single investment for most people. How can you be sure they will buy your product? Usually when correction is a way of life for a company, standard work goes out the window. What is standard work? We will get to that a little later in this book. When we fail at first-time quality, we set ourselves up for shortcuts, increased pressure, and usually longer work hours. When correction occurs, stability is compromised, which, in turn, dis-rupts the standard work of the EHS department.

When I facilitate safety and health training at a facility and the training session generates many confused questions, I view it as waste. It is obvious that I have not fulfilled my customers' requirements and that most likely I will need to rework the training and correct my employees.

OVERPRODUCTION WASTE

Waste in overproduction is considered to be the most identifiable waste. I have a surplus of cars, SKUs, or widgets and no customers. This is part of the issue with overproduction waste when you produce more than needed, but what about produc-ing faster than needed? Remember, in lean enterprise we are trying to cut the lead time down from paying for raw materials to getting paid for a finished product. If a facility produces a product faster than needed, you have equity tied up in raw materi-als, production costs, labor, and storage. The longer a finished product sits on-site, the more it will cost. The longer a finished product sits on-site, more hands will need to touch it, move it, and track it.

The elimination of overproduction waste will eliminate material handling issues associated with safety and health. Manual material handling as well as powered material handling incidents increase with overproduction as more and more inven-tory is moved and stored. Looking at it from a geographical footprint point of view, workplaces become cluttered with inventory and finished goods, which can also lead to housekeeping issues.

Of all waste that can be created in a process, overproduction is the worst type. If overproduction is occurring, it will generate almost every other type of waste identi-fied in this book. If a facility is overproducing work in progress (WIP) or finished goods, someone will need to move either WIP or finished goods around until the customer is ready. This will create movement of material and movement of operator waste. Overproduction obviously creates excess inventory and the need to track and ship excess inventory will also lead to processing waste. If overproduction is identi-fied as a waste, it should be targeted first and eliminated.

MOVEMENT OF MATERIAL WASTE

By classic definition in lean enterprise, waste in movement of material is considered any material movement that does not support a lean system. Do you as a customer care how a product was moved through a production line? Would you be willing to pay for the movement of raw material that is in your product? Movement of material is non-value-added work that must occur. In many facilities with automation, movement of material is either through powered conveyors or in some cases robotic vehicles like automatically guided vehicles (AGVs). Movement of materials is also where we start to see interfaces with machines and humans as well as a move to sequence process steps.

Elimination of any type of movement is usually a win-win for safety and health. Eliminating a material movement reduces the interface and likelihood of injuries regardless. Whether one is using a powered industrial truck (PIT) or performing a manual lift, the reduction of risk has occurred. The goal is to eliminate all waste in movement of material. Many EHS professionals see the elimination of a manual move replaced with some type of powered equipment as a win. This might be true in some cases, but your job is to try to eliminate movement altogether. Continue to push for engineering controls to eliminate employee interface with manual lifts as well as powered industrial trucks.

As an example of identifying safety issues while eliminating waste, during a kaizen event at a fabrication shop, the kaizen team identified the need to increase floor space within the department. Large metal carts were targeted for improvement and eliminated from the process. The team created new carts, reducing the footprint of the materials cart on the shop floor. The payoff for safety was a smaller cart as well. Although the goal was to reduce the footprint of the cart, in doing so the kaizen team eliminated over 120 lb. from the cart itself. In the quest to continuously improve the process, safety benefited by the reduction of force needed to physically move the cart. Figure 3.4 shows the old cart used in the fabrication, whereas Figure 3.5 shows the new cart moving the same material.

MOTION OF OPERATORS WASTE

When we look at motion in a process, we need to understand what adds value to the product or service being manufactured or created. Any movement of people or machines that does not contribute to adding value is considered waste. Time studies are normally conducted to identify the movement or motion of operators and machines to develop what is called *cycle time*. Cycle time is the time it takes to complete one step of one operation. Process steps with a large amount of motion will produce higher cycle times. There is another type of time that can be tracked called *takt time*. *Takt*, by definition, is time needed to meet customer demand. Takt time is completely different from cycle time; never confuse the two.

In one kaizen event, it was determined that employees needed to turn a pipe for welding as part of a process. The employees were asked how many pipes needed to be turned during a normal workday. The equipment used to turn the pipe was in a different work cell, further down the assembly line. It was calculated that employees were walking to turn pipes an average of 40 miles per year, at the current rate of assembly. At that point, it was determined to either move the work closer to the machine or move the machine closer to the work cell. In the end, the machine was moved closer to the employees.

Eliminating non-value-added movement is one of the biggest opportunities that safety and health has for creating change in the workplace. Identification of how operators move, and why, gives the safety program ammunition to introduce

FIGURE 3.4 A cart used to transport fabricated material before a kaizen event.

FIGURE 3.5 The same fabrication cart after the lean event. Significant weight was eliminated from the cart, reducing the force needed to move material.

ergonomic fixes in the workplace. Fixes can include moving parts from one shelf to another or creating a smaller parts bin. Many safety people get hung up on creating ergonomic fixes that include high-dollar tooling or decking equipment. Stick to the fundamentals of safety and ergonomics to make the task friendlier to the operator. Remember, kaizen events are looking for quick improvements and wins. The example of changing the fabrication cart reduced ergonomic stress quickly with relatively low cost. Go back to the basics.

During a kaizen event in an assembly area, a waste of movement was identified in the pump building area. The cart in which brass fittings were placed for eventual assembly was haphazard at best. A kaizen team was formed to minimize if not eliminate the movement of the valves as much as possible. All of the valves were brass and were assembled onto a manifold as part of a pump system. Some of the valves weighed as much as 50 lb. depending on the size of the manifold and pump application.

A cart was developed not only to hold the valves needed to build the manifold but also to hold the manifold in place during the build process. The cart could now be loaded with the proper valves for that build and the operator needed only to reach down to retrieve a valve. Figure 3.6 shows the old valve cart, and Figure 3.7 shows the new valve/manifold build cart. On the new cart, a pipe attachment was added to bolt the manifold to the cart. As the operator builds the manifold, the non-value-added movement of the manifold has been eliminated. In addition, the work zone area for the employee was raised slightly to accommodate proper work height for the task.

FIGURE 3.6　Manifold cart before a lean event.

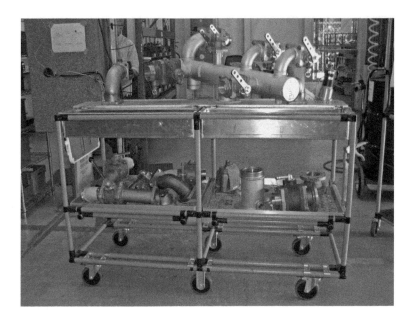

FIGURE 3.7 The same process with a new manifold cart. The new cart provided the operator a platform to build the manifold without the need to constantly rotate the manifold during the build process.

WAITING WASTE

If you are in line at a counter in a store and the employee doesn't know how to service you, have you wasted your time? There are several different types of waste for waiting, and at some point or another in your life, you have experienced all forms of waiting waste. If you have children, you have experienced every form of waiting waste this week. When we look at a waiting waste, it could possibly be a good thing for safety and health. If a process or job task is being reviewed for waste with a high frequency of operator motion, waiting can help the operator rest and "recover" until the next sequence or cycle starts. Elimination of waiting waste will also generate the biggest complaints from employees regarding safety and health. Remove the time of waiting and employees will feel like they are being rushed and forced to work unsafely. The fact of the matter is that not everyone likes change. Some employees might look at this step as a takeaway. Helping employees through change is one more reason for safety and health to be involved in lean changes in your organization. Know and understand your processes.

INVENTORY WASTE

Inventory waste is defined as any supply in excess of process requirements necessary to produce a good or service Just-in-Time (JIT). Many organizations strive to move all production parts and some services to a vendor-based inventory management system. What this means is that the vendor or contractor supplying the part manages the

FIGURE 3.8   One area in a process with excess inventory.

FIGURE 3.9   The same area after a lean event eliminated excess inventory and established Just-in-Time (JIT) selection and inventory of materials.

inventory of the stocked part or service for the end user. These management systems can be as complex as the vendor establishing a warehouse or building a facility next to a customer's facility to feed parts. The same process can be as simple as installing a vending machine for small equipment. When we look at reduction in inventory waste, we are looking to cut the time and cash spent between buying inventory to produce a product or service and getting paid for that product or service. Any time you can reduce inventory waste, you are reducing the exposure of handling parts or finished goods. Any chance to eliminate material handling is an improvement in safety and health.

PROCESSING WASTE

Processing waste is, by most accounts, the hardest waste to identify. Processing can include any effort expended on a product or service that does not add value. If you think about processing warranty claims or simple business transactions, it is transparent to the customer and sometimes even the company. Processing in most cases is non-value-added work, but necessary. The main reason why processing waste is the hardest to detect is due to the fact that you usually can't see it or feel it. Most waste elimination is visual in lean manufacturing. If I reduce inventory and change the motion of some operators, I can physically see it. Now, if I change how an engineering department releases orders to manufacturing and reduce the time by 20%, what can I see? Does the customer care that their order was processed 20% faster? No, it adds no value to the customer.

Eliminating Processing Waste—Workers' Compensation Claims

One example I use to show process waste in safety and health is the processing of workers' compensation claims. Depending on how your claims are reported, it usually requires a large amount of information to be entered into a claims system. When I went to work for a company that was self-insured, I inherited all open claims. When I looked at most of the claims from specific areas, I noticed that 90% were treated by orthopedic doctors for repetitive hand and wrist injuries. I called the third-party administrator (TPA) to determine why these cases were open so long and all with orthopedic specialists. I was told that the company doctor was told in the past to use conservative treatment for sprains and strains, especially repetitive motion. By the time the employee actually got to see a specialist, it usually meant surgery, lost time, and physical therapy. I asked the orthopedic doctor, "If we have employees with these same injuries, can we send them directly to you with a referral?" The doctor agreed, and moving forward, all repetitive motion injuries were sent directly to a specialist.

The issue came up through management that safety and health was not qualified to override a doctor, due to lack of medical training. Granted, I did not have a medical degree, but I did have common sense. The orthopedist was willing to try, the TPA liked the idea since it reduced the reserves needed to be set, and the employee was getting proper medical treatment. We reduced the processing time on medical treatment for employees with work-related injuries and reduced cost.

FIGURE 3.10   Excess inventory.

FIGURE 3.11   5S standards applied to a material storage area where excess was identified.

Intellectual Waste—The Eighth Waste

Over the past few years, many in lean enterprise have identified an eighth waste called *intellectual waste*. Intellectual waste is when a manager or company fails to identify the knowledge or ability of employees. At some point in their career, everyone has felt that their talents and ideas were underutilized. Everyone can identify a coworker who may fall into this category. Good managers and companies will consistently challenge good employees.

WASTE IDENTIFICATION AND EHS

One of the biggest tools that environmental health and safety professionals can use to help facilitate a safety program is to incorporate safety into the waste identification process. As stated before, ergonomic awareness is usually the biggest payoff for safety when identifying waste. During training exercises for kaizen team members or for kaizen events themselves, it is imperative that safety and health is identified in every process step. The sample waste identification and safety observation sheet in Figure 3.12 demonstrates how all kaizen team members can identify ergonomic stressors when identifying process waste. The identification of employee posture, the amount of force used to perform a task, and the frequency can be documented and reviewed. Manual material handling and manual lifting should also be identified and targeted for elimination. Job tasks that require fine hand manipulation or pinching as well as manual pushing and pulling are often overlooked. Identification of these ergonomic stressors and job hazards may need additional training by the safety and health professional for the kaizen team to identify.

To also help create standardized work at a standard workplace, questions can also be asked regarding job safety analysis (JSA) documentation for job tasks being reviewed during the kaizen event. If the kaizen also involves man and machine interface, this is an opportunity for the kaizen team to identify an audit lockout-tagout procedure. In Figure 3.12, a waste identification sheet has been merged with hazard recognition.

The Environmental Protection Agency (EPA) has identified key attributes and wastes that facilities and companies can identify during kaizen events. Environmental waste at the kaizen team level can sometimes be overlooked if the team is not trained properly. Some questions that the kaizen team should ask regarding environmental waste include the following:

Water Use
- How much water is used in the process and how is it used?
- How can you reuse water and/or reduce overall water use?
- Can you reduce contaminants in wastewater discharges?

Energy Use
- How much energy is used in the process and how is it used?
- How can you reduce overall energy use?
- Is equipment running or are lights on when not being used?
- Are you using efficient light bulbs?

Waste Identification & Safety Observation Sheet

Process Observed: _____ Date: _____ Observer: _____

#	Process Step	Description	Correction	Overproduction	Material Movement	Operator Motion	Waiting	Inventory	Processing	Priority
1										1 2 3
2										1 2 3
3										1 2 3
4										1 2 3
5										1 2 3
6										1 2 3
7										1 2 3
8										1 2 3
9										1 2 3
10										1 2 3
11										1 2 3
12										1 2 3

					Ergonomic Review				
#	Process Step	Description	Pinch	Pull/Push	Frequency	Posture	Force	Manual Lift?, What to Where	Priority
									1 2 3
									1 2 3
									1 2 3
									1 2 3
									1 2 3
									1 2 3
									1 2 3
									1 2 3
									1 2 3
									1 2 3

Safety Review		
Is there a Job Safety Analysis, JSA posted for this job task? □ Yes □ No	If yes, date of JSA:	Reserved
Is there a lockout/tagout posting for this job task? □ Yes □ No	If yes, date of posting:	
Reserved		

FIGURE 3.12 Waste reduction checklist with safety observations built into the checklist. Merging waste reduction with a hazard observation will help jump-start safety initiatives.

- Can you save energy by consolidating operations and/or storage space?
- Can you shift to a cleaner source of energy?

Chemicals and Materials Use
- What types and quantities of chemicals/materials are used in the process?
- How can you reduce the overall amount of chemicals and materials used?
- Can you switch to less harmful chemicals?
- Can you eliminate any non-value-added use of chemicals or materials from the product or process (excess packaging, unneeded painting, etc.)?

Solid Waste
- What types and quantities of solid waste are generated by the process?
- How can you reduce the overall amount of solid waste generated?
- How can you reuse or recycle solid wastes?
- Is there a local composting facility that the waste can be taken to?

Hazardous Waste
- What types and quantities of hazardous waste are generated by the process?
- How can you reduce the amount or toxicity of hazardous waste generated?
- Can you better isolate and separate hazardous wastes from other wastes?

Air Emissions
- What types and amounts of air emissions are generated by the process?
- How can you reduce the overall amount or toxicity of air emissions?
- How far did vehicles travel to deliver parts and supplies?
- Can you reduce the vehicle miles traveled and emissions from transportation?

There are all kinds of wastes associated with producing a product or service. Safety and health professionals need to be involved in determining waste identification to bolster the safety program. Identification of waste can be your biggest ally when implementing ergonomic fixes. Don't get hung up on expensive tooling equipment; go back to basics. Talk to the employees and operators. Make sure you understand the process and the environmental health and safety issues associated with their job tasks.

VISUAL MANAGEMENT

Visual management looks like the backbone of every safety program and standard in the United States. If you go through and look at a facility, count how many signs and visual aids you see for safety. In most cases, OSHA standards call for specific signage for different hazards, as does the EPA.

A good way to look at visual management in lean enterprise is information exchange. I should be able to walk into a facility or work cell without talking to anyone and answer the following questions:

1. What process or job tasks are being completed in the area?
2. What are the current constraints for this area (current state)?
3. What are areas for improvement and where is value being added to the process (future state)?
4. Is the facility sustaining a lean culture?

In Figure 3.13, a flag system is used as part of the visual management program. The flag system demonstrates how the product and process is working in a production area. Each flag is colored differently, depending on any issues that might be afoot in the process. Supervisors and managers can easily and visually see what is going on in the process. Colored flags mean different things:

- Green flag—No assistance needed.
- Yellow flag—Supervisor needed.
- Black flag—Work cell or employee is missing materials or parts needed to complete the job task.
- Blue flag—There is an engineering issue with the job task.
- Red flag—Work cell or employee is behind takt time. This means that the product or service will not be in time to meet customer demand.

FIGURE 3.13 A flag system is one example of a visual factory queue for managers and supervisors.

Visual factory can take on many forms in different companies and facilities. The use of flag systems, Andon lights, and other visual cues gives the employees, supervisors, and managers the ability to manage a process in a more effective manner.

A visual management system is something that all EHS departments should strive to implement. Historically, safety always has some type of visual management system in every location. Safety metrics and indicators such as OSHA injury rates and days away and restricted time (DART) rates are normal measurements for most companies and organizations. These metrics are tracked mostly because of the injury rate expectations for a specific industry set forth by OSHA. These metrics are considered by many to be *trailing indicators*, meaning that these metrics review what has already happened and do not address present or future concerns. If a facility has been tracking such things as floor safety audits or inspections and identifies a trend in unsafe conditions, corrective action could be implemented to prevent an incident. This type of safety metric is considered a *leading indicator*. Leading indicators are what lean enterprise and safety are all about. Leading indicators focus on the process; trailing indicators focus on past results.

Ford Motor Company has been well known for visual management for safety as part of the UAW-Ford partnership agreement with OSHA; Ford developed the Safety Operating System (SOS). According to the partnership agreement,

> Ford completed the development and initial launch of a new standardized safety operating system. This new system will focus on managing completion-confirmation of critical safety tasks through increased employee engagement at all levels of the local organization to ensure minimization of occupational injury-illnesses through

compliance with regulatory, contractual and corporate safety and health requirements. (Occupational Safety and Health Administration 2000)

The SOS was part of the leadership standardized work for all managers at each Ford facility. Every morning, safety and health issues were reviewed and tracked on visual boards to determine if corrective actions were needed in the safety process. Issues that were discussed included the following:

- Any injuries and illnesses from the previous day
- Training needs analysis
- Safety single point lessons (SPL)
- Review of ergonomic issues and status of ergonomic stressors

In addition, some Ford locations implemented the time spent reviewing the SOS information as a chance for supervisors and employees to present findings and corrective actions for safety and health. What the SOS does is look to the process of safety, rather than the results of safety. If an injury occurred and was reviewed at the SOS board, the supervisor or manager needed to understand the 5-Why of the incident and what corrective action would be taken to prevent recurrence. At the same time, it needed to be determined what process improvements for safety needed to be made to a system, if any.

WHAT TYPE OF VISUAL MANAGEMENT IS RIGHT?

As lean enterprise goes, so does safety. What is right for one organization might not be right for another. As companies and organizations continuously improve, different milestones are reached at different times. Lean metrics as well as safety metrics will change over time. What should be a part of your safety operating system, key performance indicators, or whatever you want to call it? The answer is—it depends.

All items and metrics tracked need to add value to the safety process. The difference with safety is that some metrics might relate back to compliance issues, and there is nothing wrong with this philosophy. Remember, OSHA standards are considered the bare minimum standard to which the government feels that employees are protected from occupational hazards. The job of the EHS professional is to create policies and programs that exceed the required recommendations to create a safe and healthy work environment. What adds value to your safety and environmental program is what you will need to figure out. All metrics should be as proactive as possible and visible for all employees. In Figure 3.14, this SOS board has identified leading indicators for safety that include the following:

- Safety audit/inspection results
- Hazard recognition/near miss reporting
- Proactive safety training completed (desired vs. required)
- Behavioral safety audits
- Pulse survey results for safety

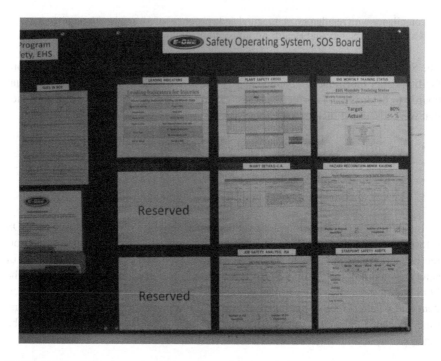

FIGURE 3.14   Safety operating system (SOS) board in a facility.

- Pre-task planning activity
- Job safety analysis (JSA) creation
- Management safety walks, GEMBA results

Different hazards, issues, and processes will call for different metrics for safety. If a gap analysis reveals that one area is in need of attention or measurement, make that the focus of the safety metrics. The metrics should add value to the safety process as well as add value to management.

5S SYSTEM

The acronym *5S* stands for five principles that have been identified by many as a foundation of lean enterprise. These principles are sometimes the first step in the lean journey and can reap heavy rewards for a safety program.

Sort—Refers to the process of keeping needed items and discarding items that are not used in the workplace. By ridding the workplace of unneeded tools and equipment, the workplace becomes less cluttered and housekeeping is improved. Figures 3.15 and 3.16 show before and after photos of the same area. By sorting what is needed, valuable floor space can be freed up.

Store—Is sometimes referred to as *set in order*. After sorting the workplace and determining what is needed and what can go, we need to figure out

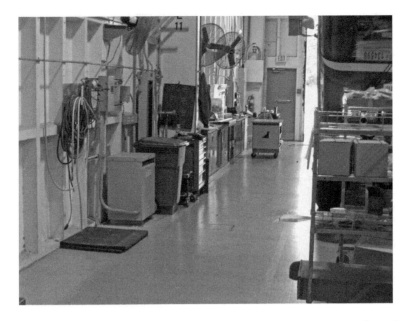

FIGURE 3.15   Sorting material before a lean event for 5S.

FIGURE 3.16   The same area after a lean event and 5S standards implemented.

FIGURE 3.17   Stored tools and parts will create waiting waste as employees need to search for tools.

where to put it. Reviewing what is used in a work cell and the frequency of use will determine where items are stored or located in the area. Figures 3.17 and 3.18 show how tools and equipment can be stored to eliminate the need to try to find the right tool for the right job.

Shine—After going through the process of sorting tools and equipment and storing them properly, make the place shine. New paint on the floors, ceiling, and walls can make a huge impact on employees as well as customers. If you have ever repainted a room in your house, you can appreciate the difference it can create. One company launched a new vehicle that had been retooled for a new model. The company had spent a total of $600 million replacing and upgrading equipment in the oldest facility in the company. Very few dollars, to say the least, were spent on shining up the facility, which was very apparent during an open house that included a media tour. Many reporters compared a facility from another company, showing a big contrast in facilities. One reporter compared it to "working in a penitentiary" rather than a state-of-the-art facility. In the examples shown in Figures 3.19 and 3.20, the difference after painting and shining up a workplace is dramatic. Any time visual improvements can be made to any workplace, it offers a psychological gain and makes employees feel better about the workplace. In Figure 3.21, the total efforts of a good 5S program provide a boost to employees as well as potential customers.

Standardize—Once a work area has been sorted, set in order or stored, and shined it will force standard work. It will be easier for employees, supervisors, and managers to identify what is standard for the work cell or zone.

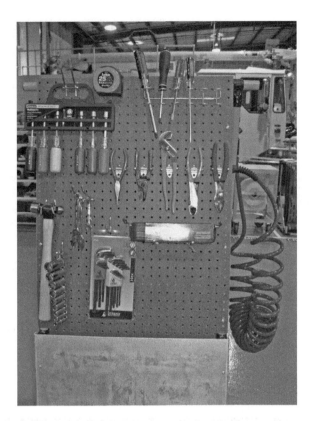

FIGURE 3.18 A-frame tool boards give all employees a visual menu to select the proper tool, eliminating the waste of searching for tools and materials.

More importantly, it will identify what is "out of place" or "not standard" for that area. Standard work also pays huge dividends for a safety program by eliminating variation in a job task. A standardized workplace is only sustained when we sort, set, and shine the workplace. Without those three principles, there will be no standardization.

Sustain—Means just what it appears to mean: sustain what you have built. The trick for many who incorporate lean into their process is the last *S* for *sustain*. Anyone who has cleaned out the junk drawer, garage, or closet will usually find themselves the following year cleaning it once again. Many companies will create 5S checklists to ensure that the workplace sustains 5S activities. It is also measurable in many lean systems.

5S Pitfalls

5S, as I stated in the beginning of this section, is sometimes the first step in a company's lean journey. It can also sometimes be the most painful. People in general will resist change, especially physical change such as removing tools and equipment that

FIGURE 3.19　An area before a lean event and 5S standards or shining the facility or area.

FIGURE 3.20　The same location, post lean event. Floors and aisleways were painted and shadow squares were placed on the floor to identify items in work cells.

FIGURE 3.21 As 5S is implemented, the psychological benefits of a clean, neat, and orderly workplace will create a workplace that employees will take pride and ownership in.

have been there longer than the employees. Many employees become extremely emotional when physical change occurs at the workplace. Change is what lean enterprise is all about: *kaizen* or "change for the better." Part of the continuous improvement process is to document real life. When we look at real life and implementing 5S, it looks more like the issues in Figure 3.22.

The problem with the kaizen blitz is that a team comes into an area and makes a change for the better. That is what the employer sees. What the employees see is a bunch of people coming into their work space and throwing their stuff away, moving their stuff to where they can't find it, and painting over a dirty floor. Same event, to very different thought processes.

In *Lean Safety*, Robert Hafey discussed the fact that "lean is 75% social and 25% technical" (Hafey 2010), meaning that communication skills have a great impact on the success of a lean event. How you communicate to the employees and people that the kaizen event touches will most likely determine the sustainability of the event. Before starting a 5S event, tell the employees what is going on—tell them before, tell them during, and tell them after the event.

Another major pitfall with 5S is the biggest pitfall of lean enterprise, failure to sustain a process. During 5S events, any one of the first three principles—sort, set, or shine—can virtually be undone within a work shift and sometimes right before your eyes. Employees will cling to tools and equipment like the *Titanic* is going down. It is up to the lean champion or sometimes the 5S champion to ensure that all efforts to 5S the workplace are sustained.

5S	Employee Perception	Real Life
Sort	I need that tool. I can't do my job without it.	This tool hasn't been used in years or the job task was moved to another work cell.
Set in Order	I like my toolbox just the way it is. I know where everything is.	The toolbox is locked and poorly sorted; no one else can get to the tools or find what they need.
Shine	Why paint the floor? It will just get dirty.	The floor hasn't been painted in the 30 years the building has been here.
Standardize	I do it this way, but Bob does it this way.	Both employees are wrong.
Sustain	This will never work.	This has to work.

FIGURE 3.22   Examples of why 5S is a "hard sell" to some employees.

How does 5S go away in a workplace after all that work? 5S creep. You have heard of creep, and we are all guilty of it. You set out to do one task, you are totally committed to it, and you end up doing something completely different. 5S creep is when we get lazy during our 5S audits and let stuff creep back into the work cell or workplace. During an audit, I noticed that 5S scores were remaining roughly the same and the place looked horrible. I talked with some of the auditors and discovered that for three different auditors, there were three different acceptance levels of 5S. How can this be? We just spent 5 months sorting, storing, and shining this place up.

As shown in Figure 3.23, new tool boards were created and old toolboxes were removed. As 5S activities for sustaining became lax, shelves were placed inside the A-frame tool boards. Slowly, old habits started to creep back into the work cell.

Like any process or system, it needs a tune-up. Any time 5S activities are sustained in a facility, you have the opportunity for 5S creep. The best solution is to get everyone back together and "recalibrate 5S eyes" to make sure that everyone is not only looking at the same things but also seeing the same things. Identify and reinforce what is acceptable and what is not.

Undoubtedly, sustainability is the hardest area of the entire 5S program to maintain. Companies' success in lean is determined by their ability to sustain the processes that have been created during the lean journey.

THE 6S ARGUMENT

Some companies have adopted the action of creating a 6S program. The additional S stands for *safety*. As part of the plan to audit the workplace for 5S, safety to some degree was added to the mix to ensure that countermeasures for safety were being addressed. There have been many arguments back and forth in the lean community regarding this additional task added to 5S due to the fact that a true lean culture builds safety into the process. This eliminates the need to address safety at this level.

Which is right for your organization and program? The answer is, it depends. Many companies will use the 6S philosophy to get a safety audit program back up and

FIGURE 3.23   Example of what happens when 5S is not sustained in the process.

running. Once 6S has been sustained for a period of time, any issues found under the safety category usually disappear from the radar screen. If a work cell or facility has been scoring high on audits for safety, does it make sense to remove the last *S*? If the culture is truly a lean enterprise, 5S will correct most, if not all, safety issues in the work cell. In Figure 3.24, a 6S checklist has been developed to include 5S plus safety.

STANDARDIZED WORK

There is no doubt that when lean enterprise is implemented in the workplace the goal is to create standard work. Standard work in the automotive industry lends itself to other areas of lean enterprise such as flow and cell design. When we eliminate non-value-added work, the safety professional can eliminate health and safety hazards at the same time. Standardizing work is the attempt to predict the outcomes of several process steps. During the identification of waste and the implementation of the 5S program, what has been created is a standard workplace. Once the standard workplace has been mapped, standard work can be identified and documented. In the automotive industry standard work is the key to stability and quality. In most cases, work that is not standard and not documented is highly scrutinized for safety and health issues.

Production - Safety & 5S Audit

Plant/Dept/Cell _____
Audit Date _____
Supervisor _____
Auditor _____

Total Score		(Total score calculation - Total Section scores, then divide by 6)
	1 = Yes / 0 = No	

Safety — "Keep clutter out of the area!"

- Are all the emergency exits clearly visible, not blocked and illuminated?
- Are all extinguishers and emergency equipment visible (red), mounted, accessible with inspections up to date?
- Are all employees wearing proper PPE (Shoes, Glasses, Ear Plugs); if welding/grinding, is additional proper gear being worn?
- Are floors free from oil, water, cords, hoses or any protrusion that may cause a fall & no daisy chaining of extension cords?
- Are electrical panels/disconnects/breaker boxes accessible & proper LOTO procedures posted & utilized?
- Are employees wearing fall protection when required and is it available where needed?
- Is personal protection equipment available & visually identified as a requirement, & is all equipment guarding in place?
- Are containers, drums, pails, bottles, etc. identified/labeled with the contents & meet our hazardous material standards?

Notes:

Score:

Sort — "Keep what is needed, throw out the rest!"

- Are only commonly used (>1/week) parts, tools and fixtures located in the area?
- Is inventory (WIP & Finished Goods) properly identified, & stored in designated areas?
- Are SQDIP, Cell Boards, Posters, signs and notices current & up to date with proper revision control documentation?
- Are work instructions and prints organized, accessible, current & contain no hand written notes for the work being done?
- Are storage cabinets eliminated from the area (storage cabinets should only be used as a last resort)?
- Are tops of shelves, cabinets and equipment free of items?
- Are unnecessary and obsolete items & scrap removed from the work area as well as no personal items in the area?
- Is red tag label information complete, current and being followed for all defective material?

Notes:

Score:

Straighten — "A place for everything & everything in it's place!"

- Are positions of main corridors, aisles, clearly marked?
- Are all material drop zones clearly marked in green and aisles clear?
- Are tool boards organized, labeled with no tools inside 'A' frame boards & are their no missing tools?
- Are all drawings, information sheets, cabinets, and shelves organized and clearly labeled?
- Are all gauges, tools, fixtures, and information boards stored and labeled?
- Is scrap and defective material seperated and controlled in the area and clearly marked?
- Are material locations labeled by part number & location number with racks clearly marked?
- Are work areas clearly marked, easily accessible, & does everything have a place & is everything in it's place?

Notes:

Score:

Shine — "Clean & check to reveal problems & improve the environment!"

- Are walls, floors and columns bright and clean (fresh paint)?
- Are the light bulbs, reflectors, top of machines, cabinets, material bins and fixtures clean and free of debris and dust?
- Are the light bulbs in operating condition (not burned out) & is their adequate lighting in the area?
- Are work surfaces and equipment clean?
- Are the SQDIP/Information boards and visual controls clean and readable?
- Are the machines free of oil, dirt and debris and been freshly painted?
- Are empty pallets, boxes and bins removed from the work areas?
- Are trash and recycle bins emptied?

Notes:

Score:

Standardize — "Apply common standards & visual management to the area!"

- Are 5S responsibilities identified and all employees trained?
- Are all storage/equipment areas marked and labeled consistently and understandably?
- Is Zone Champion Leadership Standardized Work (LSW) defined by cell and completed on time?
- Is visual management standardized (SQDIP & Cell Boards properly utilized)?
- Does the Supervisors LSW audit the success of 5S and the Zone Champions LSW?
- Is the area following the E-ONE color coding standard for all taped and painted surfaces?
- Are playbooks available, being utilized and signed-off as required?
- Is the area Flag system being utilized and the escalation process being followed for the area?

Notes:

Score:

Sustain — "Make the 5S system part of everyday life!"

- Are 5S audits done on time by the Supervisor, Cross-Audit Supervisor, and the Lean Team (minimum of 3 audits/mo.)?
- Are abnormal conditions visually and easily identifiable?
- Are 5S improvements being incorporated regularly using counter measures sheet?
- Is the 5S audit score above 6.0 for the area?
- Has progress been made on the action plan since the last review?
- Is the 5S audit score trending upward and visually posted?
- Is all quality equipment calibrated, labeled and dated properly?
- Do all documents have revision dates with current revisions being utilized for all controlled documents?

Notes:

Score:

FIGURE 3.24 Example of a 6S checklist that includes safety as a functional part of the 5S process.

STANDARD WORK AND OSHA STANDARDS

The whole idea of OSHA safety programs was the ability to create standard work or standards for safety within the United States. In the government's attempt to create standard work for safety, what was not anticipated was the interpretation of the standards. It is for this reason that many companies have different levels of safety programs to satisfy the minimum standard. In the past 10 years, OSHA has adopted a stance of issuing performance standards that allow employers the freedom to tailor safety programs to meet their standards. For example, safety standards such as process safety management (PSM) give employers the opportunity to select from a menu of different methodologies to perform process hazard analysis (PHA).

Many safety programs lend themselves to develop standard work such as lockout-tagout. OSHA 1910.147 states that all energized equipment shall have a process in place that tells the employee how to bring that piece of equipment to a zero energy state. The standard is blind to the specific equipment and the manufacturing process. The lockout-tagout standard also states that there should be a process in place to audit the lockout-tagout procedure on an annual basis. This ensures that the employer validates the process and ensures that any management of change has occurred. This form of standard work can be reviewed during a kaizen event from a safety standpoint.

JOB SAFETY ANALYSIS

Another safety tool that can be used to generate standard work is the job safety analysis (JSA). The JSA is a methodical and calculated audit of each job step in a given job task. During a kaizen event, after all the waste has been identified and non-value-added work removed, you are left with value-added work. This value-added work for this job task needs to be reviewed for safety. It is at this point that the safety professional or "competent person" must identify all hazards in the job tasks. There are several different hazard types that can be identified in a JSA. They include but are not limited to the following:

- Contact with energy sources
- Contact with hazardous materials
- Struck by moving or flying objects
- Strike against stationary or moving objects
- Caught in, on, or between anything
- Slip, trip, or fall
- Poor job design hazards such as force, frequency, or posture

In addition to the aforementioned hazard types, there are also four hazard categories, and each of these areas should be reviewed:

- Workstation
- Environment
- Personal
- Tools and equipment

When we cross-reference the hazard types with the hazard categories, safety is built into the standard work for the employee. Safety can be driven down even further in each category with the help of other lean tools such as a PPE hazard assessment. This individual assessment would fall into the personal hazard category. Although not mandatory by OSHA, the PPE hazard assessment is a standard tool to identify different types of hazards found in 1915 Subpart I, Appendix A. A company with a PPE safety program will conduct hazard assessments to identify the best PPE available. All PPE identified for the job task should be identified on a JSA. In Figure 3.25, a JSA aid has been created to help supervisors and managers identify these hazards correctly.

Job Safety Analysis, JSA Aid

Hazard	Hazard Types	Personal	Tools & Equipment	Workstation	Environment
Energy Sources	Can the employee come into contact with any energy source? • Electricity • Noise • Hydraulics • Chemicals • Compressed Air • Gravity • Heat/Cold • Gases/Radiation/Steam	• Is employee's clothing free of oil and grease? • Is jewelry absent? • Is employee knowledgeable about the energies around him/her? • Is a respirator required? • Are earplugs required? • Is any other PPE required?	• Are portable electric tools properly grounded or double insulated? • Are electrical cords in good condition? • Are air hoses in good condition? • Are tool handles insulated?	• Are tools stored correctly? • Are all machine guards in place? • Are all safety devices functional? • Are lockout instructions posted? • Are compressed gas cylinders correctly stored?	• Is lighting adequate? • Are emergency stop buttons clearly labeled? • Is heat stress a hazard? If so, is water provided? • When was the noise level last tested? • Are noise levels within acceptable limits?
Hazardous Materials	Can the employee come into contact with any hazardous materials?	• Has employee received HAZCOM training? • Does employee know where MSDS are? • Is a respirator required? • Is the respirator the correct type? • Does the respirator fit correctly? • Has the employee been trained in respirator use? • Is the respirator always cleaned after use?	• Is it possible to substitute airless spray paint guns for compressed air types? • Are screens used to enclose welding jobs? • Are tools used for application well maintained?	• Are any required types of special clothing available? Do they fit well? • Are materials correctly secured and stored? • Are materials correctly labeled? • Are separate containers provided for waste disposal? • Is waste disposed of correctly?	• Is there an effective local exhaust ventilation system? • Is there mechanical or general ventilation? • When was the air last sampled? • Is the air sampled regularly? • Is employee aware of reactivity issues?
Struck By	Can the employee be struck by anything? • Moving or flying objects • Falling material	• Is a hard hat required? • Does hard hat fit correctly? • Are safety glasses or other protective eyewear required? • Is protective eyewear of the correct type?	• Are overhead cranes maintained regularly? • Are all chains, hooks, and cables in good repair?	• Is the workstation safe from intrusion by forklifts? • Is stock correctly stacked? • Are all machine guards in place? • Are welding operations correctly guarded?	• Is lighting adequate? • Are warning bells or other warning sounds audible to the employee?
Struck Against	Can the employee strike against anything? • Stationary or moving objects • Protruding objects • Sharp or jagged edges	• Are gloves required? • Do gloves fit correctly? • Are gloves of the correct material? • Is any other special clothing required? ○ Aprons ○ Arm protection ○ Leg protection	• Are tools well maintained and unlikely to break or slip under stress? • Are tool handles correctly shaped to prevent mechanical stress?	• Is the workstation safe from intrusion by forklifts? • Is stock correctly stacked? • Are all machine guards in place? • Are welding operations correctly guarded?	• Is lighting adequate? • Are warning bells or other warning sounds audible to the employee?

Job Safety Analysis, JSA Aid

Hazard	Hazard Types	Personal	Tools & Equipment	Workstation	Environment

Caught Between	**Can the employee be caught in, on or between anything?**
	• **Pinch points**
	• **Protruding objects**
	• **Moving and/or stationary objects**
		• Does work clothing fit correctly?
		  ○ Sleeves short or buttoned?
		  ○ Pants not too long?
		• Is jewelry absent?
		  ○ Rings
		  ○ Necklaces
		  ○ Bracelets
		  ○ Loose watch
			• Are tools well maintained and unlikely to slip under stress?
			• Does the employee know the locations of emergency stop buttons and cords?
			• Are tool handles free of oil or grease?
			• Is work position a potential hazard?
				• Are all machine guards in place?
				• Does the employee know correct procedures in the event of a jam?
				• Does the employee have sufficient reach and clearance for the job?
				• Are any points of operation seven feet from the floor and below correctly guarded?
					• Is lighting adequate?
					• Are warning bells or other warning sounds audible to the employee?

Slip/Trip/Fall	**Can the employee slip, trip, or fall?** • **On the same level** • **To a lower level**	• Are safety belts or harnesses required? • Is footwear of the correct type? ○ Non-slip soles? • Does employee's clothing fit correctly?	• Are air hoses correctly positioned and away from the floor? • Are electrical cords sorted correctly? • Are tools stored correctly and away from the floor? • Are all ladders of wooden construction? • Are correct procedures used with ladders?	• Is the floor free of debris such as bolts, cardboard, spare parts, etc.? • Are platforms clean and well maintained? • Is the floor free of water, oil or other containments? • Is seating safe? • Are any pits or holes correctly guarded?	• Is lighting adequate for the job?
Ergonomics	**Can the employee be injured by poor job design hazards?** • **Frequency** • **Force** • **Stressful posture**	• Is the employee's physical condition such as to allow him/her to perform the job without the risk of whole-body fatigue? • Do required gloves fit correctly?	• Are tools the correct shape for the job? • Does the placement of tools make it easy to perform the job? • Do tools allow for alternating between hands? • Are tools well balanced? • Do heavy tools have mechanical support? • Do tools allow for neutral wrist posture?	• Does the employee have adequate room to work comfortably without being cramped or using awkward postures? • Are walking, carrying, and lifting distances within reasonable limits to minimize fatigue and potential stress? • Are displays and controls positioned for easy use?	• Does the level of lighting allow for the easy reading of displays? • Could noise level be a factor in contributing to fatigue?

FIGURE 3.25 Example of a job safety analysis (JSA) tool to help supervisors identify hazards in job tasks.

PRE-TASK ANALYSIS

Many companies believe that a JSA can only work when work is somewhat standard. The essence of the JSA is to identify each job step and communicate the hazards that accompany each step. There is another tool that is used for what can be considered nonstandard work or high-hazard work. Any department within an organization that provides maintenance, whether predictive or preventative, may engage in what is called low-frequency–high-hazard job tasks. These tasks may only be performed annually, biannually, or as needed to sustain operations. A good example of a low-frequency–high-hazard job task would be the replacement of fire protection equipment such as a water main or riser. The chances of the job task being repeated are low, but the hazard associated with the installation is high. Many times in these situations, without documenting a process of how these tasks are completed, there is an opportunity for tribal knowledge to leave the company. When we create a pre-task analysis (PTA), we create a living document for all predictive and preventative maintenance for a facility.

The PTA is a project planning tool that can be used when nonstandard work needs to be completed. Some examples of nonstandard work that can be covered with a PTA include but are not limited to the following:

- Excavation projects, depending on the depth of the dig
- Demolition/construction projects, depending on scale and scope of work
- Asbestos or lead abatement
- Projects involving the erection of scaffolding
- Projects that involve working with heights, leading-edge, or advanced fall protection systems
- Machine removal and installation

What the PTA accomplishes is to create standard work out of nonstandard work. Any time a PTA is used for a project, employees should be required to sign off on the project at the beginning of the task. Any employees added to the project would be required to review and sign off on the PTA to ensure that all employees recognize and understand all hazards associated with the project.

CELL/WORKSTATION DESIGN

During kaizen events the goal is to improve workstation and cell design at a bare minimum. This is usually accomplished through the use of the 5S program as well as the creation of standard work. Most companies that are on the lean journey see a significant improvement in safety regarding ergonomics during kaizen events. Once waste has been removed at the work cell level, such as manual material handling or excess inventory, huge improvements have been achieved.

The goal of the safety and health professional at this point would be to take the ergonomics program and the cell workstation design to the next level. If all low-hanging fruit has been picked, advanced tools for ergonomics should be included at the next kaizen event for that work cell or area. These tools can include a variety of

Project Planning Tool
Recommended to be used for High Risk Projects
And non-routine maintenance
High Risk Projects: project undertakings that are considered to be
activities outside normal maintenance process. Pre-Task Safety Analysis
supports the JSA Skill Trades and other safety processes.

Pre-Task Safety Analysis

(Form must be submitted prior to start of work to the EHS Dept)

Department:		Date:		Location:	
Main Activity/Scope of Work:					
Duration of Work:				Time of Work:	

Project Steps	Potential Hazards	Hazard Solution

Approved by Safety Eng:	Date	Field Audit by Supervisor:	Date
Approved by Supervisor:	Date	Field Audit by Safety Eng:	Date

Hourly Employee Sign-off		Supervisor/Management Sign-off	
1.		1.	
2.		2.	
3.		**Employees Assigned After Project Start-up**	
4.		1.	6.
5.		2.	7.
6.		3.	8.
7.		4.	9.
8.		5.	10.

Supervisor and Hourly Employee Sign-off: Acknowledgement must be completed before project is to be performed. Any new Supervisor and Hourly employee assigned to the project must sign the acknowledgement sheet before starting work on the project. During the course of the project the Pre-Task Safety Analysis will be periodically reviewed with all employees working on the project. The PTA process must consider sequence of work and any changes in conditions that may result in a safety hazard. An example is tear-out of a large mechanical power press where removal of the floor creates new hazards during the course of the project. PTAs must be updated to reflect these newly created fall exposures that were not included during the original PTA discussions.

Examples of High Risk Projects		
Projects	**Examples**	
Excavation Projects	Depth consideration	
Demolition/Construction Projects	Large scale	
Asbestos / Lead Work Projects	Large scale, whole plant sections	
Scaffold Projects	Height i.e. > 30', suspended scaffold	
Crane / Rigging	Large scale use, 45 ton crane	
Working at Heights	Roof decking, leading edge, Advanced fall arrest systems	
Machine Removal and Installation	Large scale, floor openings	
Other High Risk Projects identified by Plant Engineering and Plant Safety		

Examples of Projects that require Pre-Task Safety Analysis: High risk jobs often include long term projects but may also include short term projects based on their risk and complexity.

Example 1: Installation of 12-inch steam line at a plant. This project consists of installing 100 feet of 12-inch steam line. The work will be completed 35 feet off the ground using scaffolding. Fall protection / fall arrest systems, scaffold erection, rigging, etc. must be defined for the full scope of the project.

Example2: Installing an underground 16-inch fire main 50 feet long and 8 feet deep. Cave-in protection for this job needs to be defined, an excavation permit is required and the installation requires a pre-plan.

Permit / auditing systems must be used:

1. Confined Space Entry Permit   5. Electrical Distr. Permit
2. Combustion Equip Permit   6. Hot Work Permit
3. Excavation Permit   7. Local permits
4. Lockout/Tagout Audits

FIGURE 3.26 Example of a pre-task analysis (PTA) worksheet.

different assessments to calculate the current state of the work cell and what should be addressed to eliminate all ergonomic stressors. The rapid upper limb assessment (RULA) is one tool that can be used to identify work cell issues that work the upper limbs of the employees.

In many cases, safety professionals get hung up on creating an ergonomic fix that represents the ideal state of the work cell. In ideal states, money and time are no barriers and conditions are perfect. In these cases, capital improvements may be hard to justify for ergonomic issues at the cell level. Kaizen events are designed for

FIGURE 3.27   Cell design after a lean event led to elevating window glass for easy removal from dunnage.

down-and-dirty improvements and what I would consider quick wins. In Figure 3.27, a cart has been developed to hold window glass. The cart has been elevated to reduce the amount of lifting required by the employee when picking and placing glass onto the truck. In this kaizen, we have eliminated a lifting task with a lowering task.

FLOW

In lean enterprise, flow is critical in a process. Depending on what the process is, increased flow throughout a process will increase productivity. Flow can be viewed in many different ways in different processes, in both manufacturing and service industries. If you think of how freely water flows through a river, the same concept should be applied to a process. The process should flow evenly, steadily, and continuously.

Single-piece flow, or one-piece flow, is the ideal state where parts are manufactured one at a time and flow throughout a process as a single unit, transferred as customers order. Batch processing or batch flow is the opposite of single-piece flow and should be discouraged. Batch processing builds waste and non-value-added work into the process as operators or machines must sort, collect, and store batch parts for shipment or transfer to the next operator or process.

During value stream mapping (VSM), flow will be identified as either a single-piece flow or batch flow. Single-piece flow cannot be achieved every time in every process, but it should be striven for. Flow issues will also become very apparent during the VSM process. When dealing with material flow and process flow, a spaghetti chart is created. These charts are called spaghetti charts due to the fact that if flow has been ignored, once all flow has been determined as to what goes where, the chart looks like a big bowl of spaghetti. Work cells, production lines, and processes that have not been well laid out will generate huge spaghetti charts. What good process flow will do is eliminate waste in movement of materials as well as operators.

QUALITY AT THE SOURCE

Quality at the source is what drives value-based management and lean enterprise. The idea that quality is built into a process at every step will obviously produce a higher-quality product, but it will increase productivity and morale as well. Ensuring that quality is being built into every step means that all employees must adopt the policy of not doing the following:

- Creating mistakes or quality issues
- Passing on mistakes or quality issues
- Accepting mistakes or quality issues

Many companies have adopted something called the 1-10-100 Rule. This rule of thumb applies to quality at the source, meaning that if you make a mistake in a process, it will cost $1.00 for you to fix the defect. If you pass a mistake on to someone else in a process, it will cost $10.00 to fix the same defect. The price goes up as non-value-added work or waste is added to the process of correcting the defect such as stopping a line or retrieving parts or materials for correction. If a defect is made, is passed through the system, and gets out to the customer, it will cost $100.00 to correct the defect. This is only the cost of correction, not taking into account the subjective thought process of the customer buying an inferior product or service.

Several other ideas can be linked to quality at the source, such as a process called *poka-yoke* or the process of error proofing. To poka-yoke a product is to make it assemble only one way. Anyone who has tried to assemble something at home has surely struggled with parts fitting into numerous configurations. To poka-yoke would be to force assembly so it can only be built one way, the correct way.

Some considerations that should be addressed when implementing quality at the source include the following:

- Ensuring that all employees understand who their customer is and what the customer requirements are
- Awareness of quality standards, benchmarks, and processes
- Understanding the customer's intended use of the product or service
- Versatility in the workforce to provide support and help in different process steps
- Tracking, identification, and corrective action of all quality issues

If a process includes many parts from different vendors or suppliers, quality can be pushed back to the vendor/supplier as a part of incoming quality standards set by the company or organization. As parts or services are delivered to the facility, the quality of all inbound parts, assemblies, and raw material should be checked to ensure that no defects will be built into a product or service. Companies have set high incoming quality standards to the point that if a part being delivered isn't packaged or shipped correctly, the shipment is flagged for quality. The level of quality being driven will depend on what issues have been identified for a specific process, product, or service.

SINGLE-MINUTE EXCHANGE OF DYE

Single-minute exchange of dye (SMED) is a term used in lean enterprise to describe the process of changing over a production line to run a different product. The term *dye* originated from the changing of stamping dyes in power presses. One of the reasons why it was difficult to increase the variety of automobiles in the early years was the difference in stamping dyes. Dyes were very expensive to create and were difficult to change. As technology advanced, it became easier to change dyes out and have a machine back up and running, stamping a new body part or product.

SMED now also refers to any machine that needs to be changed out to run different material or a different configuration. At General Motors' Oshawa, Ontario, facility, the dyes used to stamp out the fifth-generation Camaro can be changed out in less than 2 minutes to minimize downtime in their stamping operation. As part of lean, all machines should be reviewed to identify potential changeover issues to eliminate waiting waste.

FLEXIBILITY

When we talk about flexibility in lean, what we mean to talk about is a manufacturer's ability to have a flexible process to deliver different products to the same or different customer as needed. Companies that launch new products into the marketplace hope to create a product that will be in demand. At the same time, these same companies do not want to invest capital in creating a new production line or service. The ability for one process to make several different products has become known as flexible manufacturing or flex plants.

Ford Motor Company's Chicago assembly plant became Ford's first flex assembly plant in the United States in 2004. Ford launched three different vehicles assembled at the Chicago facility, built off one vehicle platform. The Ford Five Hundred, Ford Freestyle, and Mercury Montego all had different sheet metal for the body design, but underneath all the vehicles had a standard platform. This allowed manufacturing to respond to the marketplace and customer voice without interrupting the flow of the product through the assembly line.

Many people confuse flexibility in lean enterprise with versatility. When versatility issues are determined, it usually refers to the versatility of the employees performing different job tasks. Versatility training refers to how many people can perform different tasks in a process. Depending on how lean manpower is in a given process, versatility of the employees can be the key to success.

KANBAN/PULL SYSTEMS

Kanban refers to two Japanese words, *Kan* (visual) and *Ban* (card or board). The process of kanban was fine-tuned and established as part of the Toyota Production System (TPS). During a visit to the United States, members of Toyota were benchmarking different processes in auto manufacturing and determined that many processes incorporated large amounts of non-value-added work and waste.

While in the United States, a visit was made to a supermarket. The Japanese visitors were very impressed with the presentation of the products as well as the replenishment system. All products available to be purchased were visualized. If something was selling well, it was very apparent that the supply needed to be adjusted to keep up with customer demand. Taiichi Ohno, one of the founders of TPS, brought this idea back to the machine shops in Japan.

This was the birth of what is now called kanban systems of Just-in-Time (JIT) delivery. The idea is that there is a signal card or board to identify what materials or parts were needed. The card was taken and the order filled—no more, no less. JIT eliminated overproduction and movement waste from the process.

In today's market, many companies have adopted a vendor-managed inventory system. Large manufacturing facilities have forced vendors and suppliers to create local distribution centers and mini-assembly plants to feed large facilities. The vendor or supplier is responsible for delivering exactly what is needed, when the customer asks for it. Kanban visual needs can be as simple as an empty square designated in an area, such as in Figure 3.28. The empty square tells us that whatever was there before needs to be replaced.

TOTAL PRODUCTIVE MAINTENANCE

Total productive maintenance (TPM) describes a process of engaging all levels of the organization to help increase the effectiveness of production equipment. In most organizations, maintenance has the traditional responsibility of preventive maintenance. TPM seeks to involve workers in all departments and levels, from the plant floor to senior management, to ensure effective equipment operation.

Autonomous maintenance, a key aspect of TPM, trains and focuses workers to take care of the equipment and machines with which they work. Employees operating equipment on a daily basis know and understand the equipment very well. TPM addresses the entire production system life cycle and builds a solid, plant-floor-based system to prevent injuries, defects, and breakdowns. The goal is the total elimination of all losses, including breakdowns, equipment setup and adjustment losses, idling and minor stoppages, reduced speed, defects and rework, spills and process upset conditions, and startup losses. The ultimate goals of TPM are zero equipment breakdowns and zero product defects, which lead to improved utilization of production assets and plant capacity.

In TPM, a certain amount of basic maintenance is pushed down to the employee, such as adding fluids, changing filters, and general maintenance of operations. Maintenance issues that require skilled trades such as electrical and mechanical

FIGURE 3.28   Example of a kanban square.

work are usually completed by a member of the maintenance staff. A good example of TPS is how different companies handle forklift checklists. Some companies allow forklift operators to maintain their own trucks, calling for maintenance only when the truck is due for a mechanical inspection by the manufacturer or the forklift is down. All other maintenance issues are handled from the operator perspective.

WORKS CITED

Hafey, Robert B. *Lean Safety.* New York: Taylor & Francis Group, 2010.
Occupational Safety and Health Administration. *United Auto, Aerospace, and Agricultural Implement Workers (UAW)/Ford Motor Company/ACH-LLC (#97).* August 1, 2000. http://www.osha.gov/dcsp/partnerships/national/uaw_ford/uaw_ford.html (accessed December 9, 2011).

4 Case Studies in Lean Enterprise

A bad system will beat a good person every time.

—W. Edwards Deming

E-ONE, INC., SAFETY AND HEALTH IMPROVEMENTS

E-ONE is a worldwide designer, manufacturer, and marketer of fire rescue vehicles with more than 23,000 vehicles in operation around the world. Headquartered in Ocala, Florida, E-ONE is the industry leader in product innovations, new technologies, and exceeding customer expectations. E-ONE manufactures custom and commercial pumpers and tankers, aerial ladders and platforms, rescues of all sizes, quick-attack units, industrial trucks, and aircraft rescue firefighting vehicles to meet the needs of fire departments, rescue/EMS squads, airports, Homeland Security agencies, and the military.

Established in 1974, E-ONE has grown to become an industry leader in just a few decades, and today employs more than 800 people in four plants totaling more than 420,000 square feet. Innovation has been the company's driving force and continues to be the impetus behind its pursuit of new technologies. The result is state-of-the-art fire rescue vehicles recognized for superior firefighting and rescue capabilities.

E-ONE started their lean journey in August of 2008 and continues to improve all processes within the company. E-ONE has undergone not only a visual transformation but also a cultural transformation with lean enterprise.

Over the past 3 years, E-ONE has continuously improved in all safety metrics over previous years. Major kaizen events were held to help improve processes in every department in every location. A review of 12 kaizen events occurring over the course of 1 year demonstrated some startling results. E-ONE has identified two different types of kaizen events: major events and minor events. E-ONE considers a major kaizen when a full cross-functional team is selected for the event. Minor kaizen events are driven on a department level and may not need the total resources of a cross-functional team to achieve results.

Some of the accomplishments from kaizen events included the following:

- Reducing square footage used in all processes by 8,897 square feet.
- Reducing takt time in several product lines by 1,613 minutes.
- Reduced the amount of walking by operators to get tools, materials, or parts by 56.9 miles per year.
- Increased 5S scores by an average of 4.3 on a scale of 1–8.

- In just one event, consumables were reduced by $7,000.00. In the same event, a seven-step process was reduced to five steps with an additional savings calculated at $5,340.00 annually.
- Identified and eliminated 13,886 items in materials and supplies.
- In five different process reviews, over 28 steps were eliminated in all processes for a savings of over $25,000.00.

As these improvements were made over the past 3 years, the safety program at E-ONE has been involved with all kaizen events. EHS has been involved to help identify and eliminate hazards from both unsafe conditions and acts. E-ONE EHS utilizes a fully integrated safety and health accident investigation and tracking system to help identify root cause and track corrective action. Six different areas were selected for review of incident data since the inception of lean at E-ONE.

Body parts being injured were reviewed to determine any trends. The following body parts trended and continue to trend down, some with extraordinary results:

- Eye injuries down 97%[*]
- Back injuries down 90%
- Leg injuries down 90%
- Finger injuries down 77%
- Shoulder injuries down 60%

Next, injuries by event were reviewed for positive trends. Injury events include what event caused the injury to occur.

- Employees caught in, under, or between objects were reduced by 72%.
- Employees falling from different elevations were reduced by 71%, while falls to the same level were reduced by 84%.
- Acute overexertion injuries that occur from manual materials handling and other operations were reduced by 88%.
- Repetitive trauma injuries were reduced by 87%.
- Employees that were struck by either tools or equipment were reduced by 87%.

Injuries by source were reviewed next. Injury source refers to one or more of the sources that led the employee to a workplace injury.

- Manual material handling sources were reduced by over 92%.
- Nonpowered hand tool sources were reduced by 68%.

---

[*] Eye injuries were down overall after 3 years, but eye injuries actually increased and spiked in 2009. Upon review of incident investigations, it was determined that due to all 5S activity, namely, "shining" the facility, areas in the facilities were cleaned for the first time in recent history. This, coupled with the use of cooling fans on production lines, increased the amount of foreign bodies in the air. The need for proper eye protection was reinforced and additional foreign material floating in the air was eliminated with the full completion of 5S activities.[*]

- Sources that involved using ergonomic force were reduced by 82%. This includes the use of hand tools to assemble and disassemble parts.
- Obstacles on shop floors that created a slip/trip/fall hazard were reduced by 90%.

Unsafe acts and conditions were also reviewed with trends in both improvements and deficiencies. In reviewing unsafe conditions, the following observations were made:

- Congested or restricted work areas as an unsafe condition were reduced by 91%.
- Ergonomically poor work methods were reduced by 89%.
- Defective or inadequate tools and equipment leading to injuries were reduced by 89%.
- Poor housekeeping and disorderly work cells were reduced by 84%.

At the same time, *inadequate design* as a cause of injuries under unsafe conditions increased by 30%. Inadequate design can include a part, material, workstation, or process that is not designed correctly and caused an injury. Further review of root cause in these cases identified that operators as well as supervisors were better equipped to identify design flaws in either a process or a part. In some cases, material presentation issues were corrected to eliminate the hazard.

When unsafe acts were reviewed, the following trends were identified:

- Employees who worked on small parts were failing to secure these parts in either a vise or similar equipment. Failure to secure unsafe acts were reduced by 94%. Each work cell was equipped with a station or vise to secure small parts during fabrication and installation.
- Employees not wearing the proper PPE were reduced by 90%. The implementation of standardized work helps identify the correct PPE to be used.
- Using the wrong tool/equipment for the job task was reduced by 90%. Standardizing the workplace with the proper tools helped eliminate this category.
- Improper manual material handling was reduced by 88%. Identifying flow issues using a kanban system eliminated unneeded and double handling of parts and materials.

The last category that was reviewed for safety improvements was the system or process failure mode. In other words, what process or program failed within the company that could have prevented the injury from occurring?

- PPE program failures were reduced by over 94%. The use of standard PPE, hazard assessments, and creation of standardized work helped eliminate PPE program failures.
- Failures for good corrective action were reduced by 92%. Standardized work cells and standardized work eliminated variation. Supervisors and

managers drive to root cause of all injuries and implement value-added corrective action.
- Standards, policies, or admin control (SPAC) needing improvement were reduced by 92%. Implementation of standard work and reinforcement of safety work practices helped reduce the need to improve standards.
- Material handling failures were reduced by over 88%.

Safety improvements are rarely documented in lean enterprise as safety, much like quality, is built into the process. To try to elaborate that one specific kaizen event benefited a safety program is tough to do. Lean as a whole helps identify, improve, and build safety into every process.

E-ONE, INC., PAINT PROCESS IMPROVEMENT

In January 2009, a process improvement was identified to change the paint that was used on all trucks. Several issues were identified that led E-ONE to go to a more advanced paint process. One of these reasons included the amount of hazardous waste generated from the current paint process. The team solicited several vendors for information regarding their paint process, and why their product was the best for E-ONE's applications. Several test panels were created using different paint, including destructive and nondestructive testing. After 3 months of testing, a new paint supplier was identified and brought into the management system.

As the old paint process was purged from the E-ONE system, a spike in hazardous waste was identified as old paint was discarded. Once all the old paint was removed from all sites, shipments of hazardous waste flammable liquid were reduced by over 49%. This equated to a savings of over $39,000 annually. The use of the new paint system let E-ONE mix individual colors in paint pots. This was in sharp contrast to the old system of purging and cleaning hard piped paint lines going into the paint booth. What was more interesting was what happened to the frequency of workplace injuries in the E-ONE Paint Department, specifically repetitive motion injuries.

Historically, the Paint Department incurred repetitive motion injuries over the past years due to the nature of the work. Repetitive painting, buffing, bodywork, and sanding led to cumulative trauma injuries. A trend was identified and tracked for any injuries that would occur in the Paint Department. Upon talking to several of the employees in the department, many employees stated that the new paint being used was a lot easier to work with. Employees reported that there was less buffing and wet sanding in the daily work since E-ONE started using the new paint. The Paint Department continued the incident trend and in December 2010 held a lunch appreciation event for working 1 year with no lost-time injuries.

ENVIRONMENTAL PROTECTION AGENCY CASE STUDIES

The Environmental Protection Agency (EPA) has established various case studies regarding the reduction of environmental waste at different facilities and companies throughout the United States. All of these kaizen events increased environmental awareness and established environmental green thinking into lean enterprise (U.S. Environmental Protection Agency 2011).

General Motors (GM)

General Motors Corporation (GM) has one of the most widespread lean manufacturing initiatives in place in the United States. GM grew interested in lean manufacturing in the early 1980s, as it examined elements of the Toyota Production System that had been adopted by several Japanese auto manufacturers.

In 1994, GM and Toyota formed a joint venture called New United Motor Manufacturing Inc. (NUMMI) to pioneer implementation of lean methods at an automotive manufacturing plant in the United States. Compared to a conventional GM plant, NUMMI was able to cut assembly hours per car from 31 to 19 and assembly defects per 100 cars from 135 to 45. By the early 1990s, the success of NUMMI, among other factors, made it increasingly clear that lean manufacturing offers potent productivity, product quality, and profitability advantages over traditional mass production, batch-and-queue systems. By 1997, the "big three" U.S. auto manufacturers indicated that they intended to implement their own lean systems across all of their manufacturing operations.

Since the early 1990s, GM has worked actively to integrate lean manufacturing and environmental initiatives through its Purchased Input Concept Optimization (PICOS) Program (described below). In addition, GM's WE CARE (Waste Elimination and Cost Awareness Reward Everyone) Program complements lean implementation efforts at GM facilities, as many projects result in both operational and environment improvements. The WE CARE Program is a corporate initiative that formalizes Design for the Environment and Pollution Prevention efforts into a team-oriented approach.

Saturn Kanban Implementation

Saturn's Spring Hill, Tennessee, automotive manufacturing plant received more than 95% of its parts in reusable containers. Many of these reusable containers also serve as kanban, or signals for when more parts are needed in a particular process area. This kanban-type system eliminated tons of packaging waste each year and reduced the space, cost, and energy needs of managing such waste.

Saturn had also implemented electronic kanban with some suppliers, enabling the suppliers to deliver components Just-in-Time for assembly. For example, seating systems were delivered to the shop floor in the order in which they would have been installed. Saturn also found that improved first-time quality and operational improvements linked to lean production systems reduced paint solvent usage by 270 tons between 1995 and 1996.

Fairfax Assembly Paint Booth Cleaning

At GM's Fairfax assembly plant, paint booths were originally cleaned every other day (after production) to prevent stray drops or chips of old paint from attaching to subsequent paint jobs. It was discovered, however, that the automated section of the painting operations really only needed to be cleaned once a week, as long as the cleaning was thorough, and larger holes were cut in the floor grating to allow for thicker paint accumulations. The reduction in cleaning frequency facilitates improvements in the process uptime and flow. As an additional benefit, through this and other more efficient cleaning techniques, use of purge solvent decreased by 3/8 of a gallon per vehicle. When combined with reductions achieved by solvent recycling, VOC emissions from purge solvent were reduced by 369 tons in the first year following these adjustments.

APPLICATION OF LEAN METHODS TO ADMINISTRATIVE PROCESSING IN THE PURCHASING GROUP

In addition to applying lean thinking to manufacturing processes, GM has looked at ways to lean its internal administrative processes. For example, GM's purchasing group investigated the company's request for quote (RFQ) processes by which supplier products are sought. Because each RFQ has to include a detailed listing of system requirements, RFQs under the prior paper-based system could be quite large, ranging in size (in total paper "thickness") from 3/4 of an inch to 6 inches thick.

Upon applying a value stream mapping and analysis, GM identified a number of ways in which this process produced an excessive amount of waste. Not only did it require GM to purchase and use a great deal of paper, but it also incurred costs and used raw materials associated with printing and packaging, in addition to cost and energy required to deliver each package to each potential supplier.

GM's solution was to transform the RFQ process into an electronic-based system that not only is paperless but also avoids the additional costs and waste associated with printing, packaging, and shipping each RFQ. Using an Internet-based system called Covisint, GM is able to improve procurement efficiency while lowering costs by saving time and eliminating waste. By distributing RFQs electronically, GM estimates that the company will save at least 2 tons of paper each year.

LEAN ENTERPRISE SUPPLY CHAIN DEVELOPMENT

In the early 1990s, GM realized that it was not sufficient to just lean GM's operations, as GM (and the customer) directly bears the costs of supplier waste, inefficiency, delays, and defects. GM assigned a group of engineers to work more closely with its suppliers to reduce costs and to improve product quality and on-time delivery.

This effort has involved over 150 supplier development engineers conducting lean implementation workshops called Purchased Input Concept Optimization with Suppliers (PICOS). As part of PICOS, small teams of GM engineers visited GM suppliers for several days to conduct training on lean methods and to lead a focused kaizen-type rapid improvement event. Follow-up was conducted with the suppliers at 3 and 6 months to determine if productivity improvements had been maintained, and to assist with additional process fine-tuning.

Over time, GM found that having an engineer involved in the PICOS program who is familiar with environmental management provided important benefits for leveraging additional environmental improvement from the PICOS events. By working with suppliers on environmental improvement, GM has also, among many things, been able to do the following:

- Promote the use of returnable shipping containers in lieu of single-use, disposable ones
- Communicate GM's guidelines for designing for recyclability and broadly disseminating its list of restricted or reportable chemicals
- Communicate success stories to the supplier community as examples of what can be done

GM also announced that suppliers will be required to certify the implementation of EHS in their operations in conformance with ISO 14001. GM is currently developing a broader supply chain initiative, with involvement from EPA and NIST, that some participants hope will become a vehicle to integrate technical assistance on advanced manufacturing techniques and environmental improvement opportunities.

STEERING COLUMN SHROUD PICOS EVENT

GM conducted a PICOS rapid improvement event with a key supplier to enhance the cost competitiveness and on-time delivery of steering column components. The GM team used value stream mapping and the 5-Why process to assess the existing process for steps that cause long lead times and delays. The assessment revealed that the supplier shipped the steering column shrouds (or casings) to an outside vendor for painting prior to final assembly with the steering column, adding significant flow time to the production process.

Using the 5-Why technique, the team asked why the shrouds needed to be painted in the first place. The answer was "because the die (plastic mold) creates flaws that need to be covered." This led the team to a simpler, less wasteful solution—improve the quality of the die, and mold the part using resin of the desired color.

After some research, and capital investment of $400,000, the supplier incorporated an injection molding process for the shrouds into the assembly line, eliminating the need for the time-consuming painting step. This project saved the supplier approximately $700,000 per year, while shortening lead times and improving on-time delivery to GM.

This lean project produced environmental benefits, although they were not needed to make the business case for pursuing the project. Elimination of the painting process step eliminated the following:

- 7 tons per year of VOC emissions from the painting process step
- All hazardous wastes associated with the painting process step (including cleanup rags, overspray sludge, off-spec and expired paints)
- Environmental impacts associated with transporting the shrouds to the painting vendor and back

THERMOPLASTIC COLOR PURGING PICOS EVENT

While working with a supplier to reduce lead times and improve quality for the production of a thermoplastic molded component, a GM-facilitated team found additional waste elimination opportunities associated with color changeovers. At this time, the suppliers' operations were running 7 days a week to meet customer demand.

The team found that each time the supplier changed resin colors to produce a new batch of parts, as many as 5 to 10 large plastic parts needed to be scrapped. The accumulated scrap typically would fill a 30-yard dumpster every day, resulting in $3,000 to $4,000 per week in disposal costs. In addition, the supplier consumed more resin than necessary, contributing to higher material costs.

By focusing the rapid improvement event on streamlining the die setup and color changeover process, the team was able to reduce the need to run overtime shifts to meet customer demand while eliminating a significant waste stream, as well as the extra resin and processing associated with the scrap.

LOCKHEED MARTIN—LEANING CHEMICAL AND HAZARDOUS WASTE MANAGEMENT

In 1995, the Lockheed Martin Manassas plant's CESH department conducted improvement events to apply lean thinking to its chemical and waste management activities. The key drivers for this initiative were to significantly reduce the cost, space, and staffing needed to support chemical and waste management activities at the plant. These were critical needs since Lockheed Martin divested itself of one of two semiconductor manufacturing operations at the Manassas plant. The one kept is smaller in scope, and it focuses on research and development rather than production. In their light manufacturing operations, semiconductors needed for production are purchased from off-site suppliers.

Prior to the lean event, chemical management at the facility focused on a chemical storage warehouse (64,000 square feet) containing a large buffer inventory of chemicals to ensure 100% availability. Chemicals were typically ordered quarterly in larger volumes under a blanket purchase agreement. Chemicals were stored in the warehouse until withdrawn by operations. Lockheed Martin found that a significant portion of warehoused chemicals was going directly to the hazardous waste stream without ever being used, when they expired on-shelf or when they were no longer required for research or production. Prior to the lean event, hazardous waste management activities at the plant were governed by an RCRA Part B permit.

The lean event aimed to move toward a Just-in-Time chemical management system, where chemicals are delivered three times each week in "right-sized" containers to meet real-time demand (influenced by prior week consumption rates). The objective was to dramatically reduce chemical inventories, except for selected specialty chemicals with longer lead times for acquisition and delivery. Several lean principles guided the events: (1) optimize performance for the entire system even if per-unit chemical purchase or waste disposal costs increase, (2) focus on actual needs, not worse-case contingencies, and (3) focus on smooth flow of materials through the facility.

The new system also eliminated the chemical warehouse, replacing it with point-of-use storage (POUS) cabinets and right-sized containers of chemical supplies.

Lockheed Martin has contracted with five to six suppliers (multiyear agreements) to deliver the chemicals to the facility's chemical handling dock. CESH staff then transport the chemicals from there to the POUS cabinets. Lockheed Martin has shifted its relationship with chemical suppliers to more of a partnership model, with provisions and incentives for ensuring prompt delivery and chemical availability, while limiting on-site inventory. The facility's Chemical Challenge Program poses questions up front, at the product and process design stage, which explore opportunities to minimize chemical usage and risk.

The lean event also sought to reduce the total waste management system cost by eliminating on-site treatment and the need for the RCRA permit, shifting to regular hazardous waste pickup by a waste management vendor. By switching from a practice that purchased and stored on-site quantities of chemicals based on estimates for the upcoming production to a purchasing practice that is driven primarily by purchasing chemicals just when needed (Just-in-Time; Point of Use [POUS]; and right sizing chemicals) they managed to slash significantly the quantities of wastes generated at the facility. They are now a 90-day RCRA–Subtitle C–Large Quantity Generator. In fact, they have several 90-day satellite storage areas. This is because they use chemicals and generate hazardous wastes at other parts of the facility in addition to the one they leaned out of a Part B permit.

A summary table is provided below that compares the prior and current methods for chemical and waste management at the Manassas plant. The lean events achieved the following business results related to the chemical and hazardous waste management processes at the Manassas facility:

- Chemical inventories were dramatically reduced, freeing capital tied up in inventory.
- Chemical inventory turns dramatically increased.
- Chemical utilization rates increased dramatically, virtually eliminating chemicals expiring on the shelf or being mixed in larger quantities than needed.
- Chemical warehouse was eliminated, reducing chemical storage space from 64,000 square feet to 1,200 square feet.
- Despite increased unit cost for hazardous waste disposal/treatment, significant system savings have resulted from elimination of the RCRA Part B permit and associated regulatory requirements.

These business results from the lean events also produced several environmental benefits:

- Reduction in chemical inventories reduces the likelihood of chemical-related spills and accidents.
- Virtually eliminated hazardous waste caused by chemicals expiring on the shelf and from excess chemicals mixed in quantities larger than needed.
- The chemical authorization process and chemical challenge program tightened screening of chemical choices and increased attentiveness to chemical use and risk reduction opportunities.
- Energy savings resulted from the significant reduction in warehouse space required for chemical storage.

TABLE 4.1

Summary of Lean Initiative Results

Activity	Prior Method	Current Method	Benefits and Concerns
Scope	Two semiconductor facilities and light manufacturing	One semiconductor facility and light manufacturing	Remaining mfg is 20% of prior scope.
Total facility size	1,650,000 square feet	1,100,000 square feet	Several buildings sold but added four small facilities in other states (NY, CA, FL).
Basis for chemical purchases	Support staff estimate based on prior use with buffer to ensure 100% availability	Order as needed based on prior week consumption and lead time for specialty items	No extra chemicals ordered. Virtually eliminated waste caused by expired shelf life and unused chemical waste.
Contract with supply	Multiyear agreements	Multiyear agreements with delivery and availability addressed	Supplier more of a partner. Minimal inventory storage shifted to supplier.
Hazardous waste	RCRA Part B Permit	Large quantity generator	Went to pick up by vendor on a schedule. Minor increase in unit cost, significant savings by eliminating permit requirements.
Chemical storage	64,000 square feet	1,200 square feet	Significant cost savings but minimal room for future growth.
Staffing	64	17	Reduced work scope and elimination of unneeded work. No backup support.
Departments	5	1	Consolidated engineering, operations, chemical, health, safety, environmental, industrial hygiene.

REJUVENATION—SUSTAINABILITY AS BASIC CORPORATE VALUE

REJUVENATION AND ITS THREE P'S

Rejuvenation remakes history. Committed to historic preservation and renovation, the Portland, Oregon, company manufactures and sells custom, period-authentic reproduction lighting and hardware for use in older homes and buildings. The company is guided by a philosophy that the reuse and improvement of old properties is desirable to encourage, which has clear environmental sustainability benefits. Last year, Rejuvenation generated nearly $35 million in sales, primarily through direct marketing and two retail stores in the Pacific Northwest, with sales growth averaging about 10% annually. The company was founded in 1977 and had approximately 240 employees in July 2006. Through its incorporation of an environmental management system (EMS) based on the Natural Step (TNS) and use of lean manufacturing practices, Rejuvenation has been able to reduce its footprint, increase quality and profit, and stay true to its core corporate values.

Rejuvenation's goal is efficient, high value-added use of resources with minimal environmental impact. Jim Kelly, founder and CEO, has always placed a high value on socially responsible business practices. From their environmental footprint to employee treatment to customer relations, he wants everything to work well. Therefore, it is not surprising that the company has made sustainability one of its core values. Indeed, the company's philosophy can be summed up by what it calls the "three P's: people, profit, planet."

THE NATURAL STEP

In 1997, Jim Kelly and John Klosterman, vice president of operations, attended a conference about TNS. TNS is a framework that uses the natural sciences to guide businesses toward more sustainable—and profitable—practices by intelligently integrating the environment into decision making and operations. Soon after the conference, Rejuvenation volunteered for an innovative pilot project sponsored by the Oregon Department of Environmental Quality (DEQ), the City of Portland Bureau of Environmental Services, and the Oregon Natural Step Network that integrated TNS with an environmental management system (EMS) based on ISO 14001. This system would allow Rejuvenation to analyze and mitigate the environmental impact of its manufacturing practices.

Now, Rejuvenation performs an annual "aspects analysis" screening to score every facet of all its product lines and manufacturing practices on their adherence to the company's core principles, always considering the scale of any environmental impact. Case in point, they examine the entire life cycle of each of their products, from suppliers' fabrication of parts through the customer's ultimate disposal of the product. Through these scores, they identify opportunities and create a prioritized list of changes for the company. For example, through this analysis, they saw that the largest impact on their carbon footprint came from customers' use of Rejuvenation incandescent light fixtures. Lumen for lumen, incandescent lighting is less efficient than other lighting sources and thus requires more electricity, which, in turn, leads to more emissions from power plants. To mitigate this, Rejuvenation increased its selection of period-authentic lighting fixtures that accommodate more efficient, and therefore less polluting, compact fluorescent (CFL) bulbs. Rejuvenation has also focused on product durability, striving to manufacture products that will last for a long time.

INTRODUCTION OF LEAN MANUFACTURING

Due to growing demand, Rejuvenation moved to a new factory in 1998. Around the same time, it dropped its craftsman-based production methods because of the need for more consistent quality and greater manufacturing efficiency. Having learned the benefits of flow manufacturing at a training workshop, Klosterman and his managers began to experiment with lean methods. To facilitate the introduction of lean practices to Rejuvenation, Klosterman hired a manager with lean experience and offered him a "blank slate" (i.e., the new factory). Today, Rejuvenation commonly uses lean tools like kaizen, VSM, 5S, waste minimization, and root cause analysis. Rejuvenation has pursued what they refer to as a "hybrid approach" to lean that adapts lean methods to their low-volume, high-variety manufacturing system. In addition, they are willing to reexamine the same operation year after year, thus allowing fresh eyes to analyze the process and identify opportunities for further improvement. Because the changes made have been continuous and gradual (kaizen events notwithstanding), it has taken some time for everyone to recognize the full potential of lean. In fact, Klosterman believes that most employees have only fully absorbed lean thinking and principles in the last year or two. He contends they are now able to do things that he could have never dreamed of just a few years ago.

Mass Customization

As a "mass customization shop," Rejuvenation produces a small volume at unit-level of a large variety of products. They strive to produce a product that meets a customer's precise needs. Prior to the introduction of lean, they sent finished products to a quality control department for inspection. Therefore, if an operator accidentally used a part or finish that did not exactly match the customer's preference, the mistake would not be caught until the end of the line. The result was a 12–15% failure rate requiring a lot of expensive rework. Through lean, Rejuvenation dissolved the department and made quality control everyone's job. Now, customization errors can be identified and corrected before the next process begins. In addition, under the theory that quality improves if workers can see and value the whole process and not just a few narrow tasks, they dismantled an individual productivity system that had prevented employee flow across departments and teams. By building quality control into the entire manufacturing line and instilling the value of quality in every employee, Rejuvenation was able to broaden its catalog and offer "Tinkertoy" customization that allows customers to choose novel assemblies of a host of styles, finishes, and accessories.

A customer order provides the initial pull. Upon receipt of an order, a kit containing the required parts and materials is assembled and then pulled through the manufacturing line using traditional lean pull signals at each step. Because it promises customers a 14-day lead time, whereas its overseas suppliers can require a 3- to 4-month lead time, Rejuvenation does maintain a significant inventory of raw materials and parts. Their inventory of parts gets completely turned over only about once a year. However, they believe this works for them. Nevertheless, through lean, work-in-progress (WIP) has been significantly reduced. Although as a customization shop, Rejuvenation has never considered WIP as much of a problem, over the last 5 years it has been cut from 7 to 8 days to about one and a half days. Interestingly, it took floor operators a while to get used to this improvement; they had been using the WIP as a signal to gauge their work speed.

Correct Use of 5S

Rejuvenation uses 5S constantly in the shop. As a result, the factory floor is much cleaner and better organized than it was 8 years ago. However, workers are allowed to vary their routines to prevent monotony and allow some limited individualization of practice. Klosterman believes that workers need some level of personalization to be fully committed to continuous improvement. He warns that 5S can sometimes become an unhealthy obsession or an end unto itself, rather than a means toward improving efficiency and reducing costs. Like all lean tools, it must be applied carefully, practically, and appropriately.

Small Changes, Big Results

Although they make no formal connections between their lean practices and sustainability efforts, they recognize that the two are connected and complement each other well. Nevertheless, they rarely make an effort to calculate the exact environmental or financial benefits from a single, small change. Instead, they view their practices as a holistic approach with aggregate outcomes (requiring aggregate measures). More precisely, they believe that a small change can trigger a chain of positive effects. They ask, "What are we doing?" on an everyday basis and then look for possibilities there. For example, they recognized an opportunity to confine a hazardous material, antiquing wastewater, by changing the antiquing process. Confining that waste stream reduced the overall discharge, which allowed the creation of a closed-loop zero-waste system in that area. This new system gave them greater control over the process and quality, which then allowed them to switch from a flocculation process to a resin bed system. This resin bed system reduced the cost of regulatory compliance and reduced water usage through recycling, which lowered their sewer charges. They believe that the cumulative positive effects of these changes could be missed if the company were too focused on measuring the relative merits of a change in one small subprocess alone.

Regulators: Partners and Valued Experts

Interestingly, Rejuvenation does not have any staff members solely dedicated exclusively to regulatory compliance. As with quality control, they believe that compliance is everyone's job. In fact, they tend to view regulatory agencies as partners rather than "cops on the beat." For example, the company has excellent relations with the Oregon DEQ and the City of Portland, both of which were integral in the implementation of the TNS-EMS. The company believes that regulatory agencies, like EPA, often have valuable expertise that can help them improve their operations and environmental performance.

Ultimately, Rejuvenation wants its investments in time and money to improve quality, reduce costs, and improve marketability—recognizing that environmental benefits often make good marketing—while adhering to the core values of the three P's. Through its use of lean practices and an EMS based on the Natural Step, Rejuvenation has a strong foundation to continue growing and improving sustainably.

The EPA has an entire website dedicated to lean manufacturing and the environment. Tools on the website include four different toolkits to help companies implement lean to reduce the environmental footprint. Toolkits include the following:

- The Environmental Professional's Guide to Lean & Six Sigma
- The Lean and Chemicals Toolkit
- The Lean and Environment Toolkit
- The Lean and Energy Toolkit

This is a tremendous resource for the EHS professional. These toolkits will help you ask the right questions and identify environmental stressors in any process that you might have at your site. Go to http://www.epa.gov/lean to learn more.

WORKS CITED

U.S. Environmental Protection Agency. *Case Studies & Best Practices.* March 31, 2011. http://www.epa.gov/lean/studies/index.htm (accessed January 3, 2011).

5 Managing Change, Stress, and Innovation

I put a dollar in one of those change machines. Nothing changed.

—George Carlin

Usually when companies move toward lean enterprise, the learning curve can be too much for some people. When we talk about kaizen, we talk about change for the better. In most organizations, change is seen as anything but good. People in general resist change due to a fear of the unknown. The best way to manage change is to create change.

The second best way to manage change is to communicate to all employees why something is changing and when. If a facility is going to implement lean enterprise, the change will be immediate as soon as 5S activities start and waste is identified. Employees will be asked to eliminate old equipment and old work practices and, more likely than not, to organize their workplace. Employees with long tenure usually resist change the most, creating more stress in the workplace. It is for this reason that up-front communication regarding the lean process be conducted before any kaizen events occur.

Kaizen events should always include employees who work in the area. Only the employees working in the process to be improved know the pitfalls and issues that need to be addressed. Involving the employees also creates significant team-building events for an organization. Kaizen teams should be made up of a cross-functional team from all areas of the business, not just one department or area. These team-building events help all employees understand the other aspects of the business from a different point of view. It also builds a personal sense of responsibility to the process being improved and the people who work in the area.

As discussed previously, the hardest part of lean enterprise is sustaining change. No one ever comes to work thinking that they want to do a bad job. The same holds true for kaizen teams. No one wants to go into a kaizen event and change the process for the worse. This is one reason why the development of standardized work is so important in lean enterprise. Developing standard work should help eliminate and address workplace stressors regarding change of a process. All value-added work has been identified and verified to cycle time, giving the employee the necessary time, information, and tools to complete the job tasks. Lean enterprise is only as good as its weakest link. Miss any steps in the process of building a lean house, especially in the foundation, and the house will not stand.

Depending on how large an EHS department is, managing change and stress during lean events can be difficult. The key to a successful kaizen from the EHS standpoint is to make sure that people have the right tools and information. If at all possible, have a member of the EHS department on the kaizen team. This is the time to complete hands-on hazard recognition training with other members of the kaizen team to identify workplace safety issues. Train kaizen and lean leaders in environmental health and safety programs and standards to ensure that any issues identified in kaizen events get proper attention. Identify EHS department contacts for kaizen teams to contact when issues are discovered or identified during the event. Develop and implement an EHS checklist for the kaizen team to use to identify safety and health issues not covered in the kaizen event. This checklist can also be used as a check and balance to identify continuous improvement issues as well as successful implementation of lean to eliminate EHS issues.

Obviously, the ideal situation would be to have a member of the EHS department on the kaizen team. There is no substitute for being part of the team to identify safety and health issues at the source. In addition to the four areas described above, a member of the EHS department should periodically check in on the kaizen team during an event. This will ensure that any issues that are related to safety and health are answered as quickly as possible. If EHS is not called at least once during a kaizen event, it would be in the best interest to check the status of the team. A team that is cross-functional throughout an organization should generate numerous safety and health questions regarding continuous improvement efforts.

In the post-kaizen checklist identified in Figure 5.1, any process changes that can impact environmental health and safety can be identified by a simple yes/no checklist. This checklist is broken down into three different categories, including ergonomics, safety, and environmental. This checklist should identify variable issues that can be impacted with any process change.

Managing stress in environmental health and safety is no different than managing stress in other jobs. Many safety professionals endure a burnout rate higher than most other professions in manufacturing. There are a number of factors contributing to this, including lack of compliance, lack of management buy-in, and individual work hours. Most safety professionals wear multiple hats that can include emergency response, occupational safety, environmental, workers' compensation, and DOT compliance. EHS departments are shrinking as the economy rebounds from its latest downturn. More companies have learned to do more with less, placing an increased stress on the current workforce. Although resources have been diminished, the question still remains as to how much work is value-added versus non-value-added for the safety professional. To help drive the safety culture and eliminate stress, every effort should be made to train other employees to identify health and safety issues during lean implementation.

TRAINING KAIZEN TEAM MEMBERS IN SAFETY AND HEALTH

One of the best tools that can be used to train employees in health and safety is the OSHA outreach training for general industry and construction businesses. These classes are commonly 10-hour or 30-hour classes and are usually given by Outreach Training Institutes (OTI). Each region within OSHA will have several OTI centers for each region. The OSHA 10-hour class for general industry covers national emphasis programs that OSHA has determined as leading indicators for occupational safety and health nationwide. Class modules are reviewed and tweaked by OSHA as needed to help reduce and eliminate workplace injuries.

Post-Kaizen EHS Checklist | 2011

The purpose of this checklist is to review any process changes that impact Environmental Health and Safety, EHS. This is also a "lessons learned" tool to help future kaizen teams identify EHS improvements. **Has the Kaizen changed any of the following areas regarding EHS?**

Area	#	Question	Yes	No	N/A	Notes/Issues
Ergonomics	1	Reduced the amount of lifting required for job tasks?				
	2	Created a Push/Pull/Carry task for the process?				
	3	Eliminated one hand lifts or team lifts?				
	4	Created a one-hand lift or team lift process?				
	5	Reduced vibration contact with powered hand tools?				
	6	Increased lowering tasks instead of creating lifting tasks?				
	7	Is ergonomic matting in place for static workstations?				
	8	Has lighting been changed and/or sufficient for job task?				
Safety	1	Moved any fire protection equipment?				
	2	Requires the use of a Powered Industrial Vehicle, PIT?				
	3	Changed egress or exit routes?				
	4	Eliminated extension cords or power strips?				
	5	Created any Job Safety Analysis, JSA documents?				
	6	Changed a Lockout/Tagout procedure?				
	7	Has piping been properly labeled with contents and flow?				
	8	Have any chemicals been eliminated or added?				
Environmental	1	Eliminate any type of energy use? Machines/lights left on?				
	2	Eliminate any water waste in the process? Process clean up, wastewater, etc.				
	3	Eliminate any type of air waste? Sanding dust, chemical evaporation				
	4	Eliminate any type of solid waste? Universal trash, scrap, etc.				
	5	Eliminate or minimize the disposal hazardous waste? Solvent, paint, degreasers, etc.				

FIGURE 5.1 Example of a post-lean event or kaizen checklist to ensure that no safety or environmental issues were created during the event.

Over the past few years, several changes have been made to the first module, which reviews the OSH Act, General Duty Clause, and other information that informs the trainee on employee rights under the OSH Act. Many trainers and consultants have received pushback from employers regarding the recent changes to the training program. Many feel that the training module is pushing a pro-labor agenda, forcing more enforcement than creating a positive hazard recognition program. The bottom line is, if you are doing what you are supposed to be doing, this is a nonissue (Occupational Safety and Health Administration 2009).

The 10-hour General Industry Outreach Training Program is intended to provide general industry workers a broad awareness on recognizing and preventing hazards on a general industry site. The training covers a variety of safety and health hazards that a worker may encounter at a general industry site. OSHA recommends this training as an orientation to occupational safety and health. Workers must receive additional training on hazards specific to their job. Training should emphasize hazard identification, avoidance, control, and prevention, not OSHA standards. Instructional time must be a minimum of 10 hours that must be covered over a 2-day period. OSHA has identified mandatory or required training in the 10-hour class as follows:

1. 1 hour—Introduction to OSHA
   a. OSH Act, General Duty Clause, Employer and Employee Rights and Responsibilities, Whistleblower Rights, Recordkeeping Basics
   b. Inspections, Citations, and Penalties
   c. Value of Safety and Health
   d. OSHA Web site and available resources
   e. OSHA 800 number
2. 1 hour—Walking and Working Surfaces, Subpart D to include fall protection
3. 1 hour—Exit Routes, Emergency Action Plans, Fire Prevention Plans, and Fire Protection, Subparts E & L
4. 1 hour—Electrical Safety, Subpart S
5. 1 hour—Personal Protective Equipment, Subpart I
6. 1 hour—Hazard Communication, Subpart Z

OSHA has also identified specific training for different industries to be covered in mandatory training:

- Medical/Health Care—1 hour each—Introduction to Industrial Hygiene, Bloodborne Pathogens
  - At least 1/2 hour each—Ergonomics and Workplace Violence
- Maintenance—Ergonomics and (if applicable) Powered Industrial Trucks
- Utility—Ergonomics, Powered Generation, and Confined Spaces
- Office—Ergonomics

Everything that has been discussed to this point has been what OSHA requires to be included in the training; several electives can be picked by the trainer or trainee to be covered as well. These electives must add up to at least 2 hours with a minimum of 30 minutes on each topic. These electives include the following:

- Hazardous Materials, Subpart H
- Materials Handling, Subpart N
- Machine Guarding, Subpart O
- Introduction to Industrial Hygiene, Subpart Z
- Bloodborne Pathogens, Subpart Z
- Ergonomics
- Safety and Health Program

The total class time with mandatory and elective classes adds up to 8 hours with the additional 2 hours allowed teaching other general industry hazards or policies and/or expanding on the mandatory or elective topics.

The 30-hour General Industry Outreach Training Program is intended to provide a variety of training to people with some safety responsibility. Workers must receive additional training on hazards specific to their job. Training should emphasize hazard identification, avoidance, control, and prevention, not OSHA standards. Instructional time must be a minimum of 30 hours.

MANDATORY TRAINING MODULES UNDER THE 30-HOUR GENERAL INDUSTRY OUTREACH TRAINING PROGRAM

1. Introduction to OSHA—at least 2 hours
   a. OSH Act, General Duty Clause, Employer and Employee Rights and Responsibilities, Whistleblower Rights, Recordkeeping Basics
   b. Inspections, Citations, and Penalties
   c. General Safety and Health Provisions, Competent Person, Subpart C
   d. Value of Safety and Health
   e. OSHA Website, OSHA 800 number, and available resources
2. Walking and Working Surfaces—including fall protection, Subpart D—at least 1 hour
3. Exit Routes, Emergency Action Plans, Fire Prevention Plans, and Fire Protection, Subparts E & L—at least 2 hours
4. Electrical, Subpart S—at least 2 hours
5. Personal Protective Equipment (PPE), Subpart I—at least 1 hour
6. Materials Handling, Subpart N—at least 2 hours
7. Hazard Communication, Subpart Z—at least 1 hour

ELECTIVE TRAINING MODULES UNDER THE 30-HOUR GENERAL INDUSTRY OUTREACH TRAINING PROGRAM

Five different topics should be identified and add up to a minimum of 10 hours of training. Electives include the following:

- Hazardous Material (Flammable and Combustible Liquids, Spray Finishing, Compressed Gases, Dipping and Coating Operations), Subpart H
- Permit-Required Confined Spaces, Subpart J
- Lockout/Tagout, Subpart J

- Machine Guarding, Subpart O
- Welding, Cutting, and Brazing, Subpart Q
- Introduction to Industrial Hygiene, Subpart Z
- Bloodborne Pathogens, Subpart Z
- Ergonomics
- Fall Protection
- Safety and Health Programs
- Powered Industrial Vehicles

Outreach Training Institutes (OTI) are located throughout the country, including territorial areas of the Unites States. Training is value-added to all organizations as well as cost-effective. Below is a list by region of all outreach training centers.

Region I

- Keene State College, OSHA Education Center, 175 Ammon Drive, Manchester, NH 03103-3308, Phone: (800) 449-6742 Fax: (603) 358-2569

Region II

- Rochester Institute of Technology, OSHA Education Center, 31 Lomb Memorial Dr., Rochester, NY 14623-5603, Phone: (866) 385-7470 x-2919 Fax: (585) 475-6292
- University at Buffalo, 3435 Main Street Room 134, Buffalo, NY 14214-3000, Phone: (716) 829-2125 Fax: (716) 829-2806
- University of Medicine & Dentistry of New Jersey, 683 Hoes Lane West, Piscataway, NJ 08854, Phone: (732) 235-9450 Fax: (732) 235-9460
- Universidad Metropolitana, PO Box 278, Bayamon, PR 00960-0278, Phone: (787) 288-1100 x-1375 Fax: (787) 288-1995

Region III

- Chesapeake Region Safety Council, 17 Governor's Court, Baltimore, MD 21244, Phone: (800) 875-4770
- ECRI Institute Headquarters, 5200 Butler Pike, Plymouth Meeting, PA 19462-1298, Phone: (877) 700-6212, Fax: (610) 834-1275
- Johns Hopkins University & Health System, 2024 E. Monument Street, Baltimore, MD 21205-2223, Phone: (877) 700-6212
- Mid-Atlantic Construction Safety Council, 1717 Arch Street, Suite 3370, Philadelphia, PA 19103, Phone: (215) 557-1961
- Center to Protect Workers' Rights – The Center for Construction Research and Training/Building Construction Trades Department AFL-CIO, 8484 Georgia Avenue, Suite 1000, Silver Spring, MD 20910-5613, Phone: (301) 578-8593 Fax: (301) 578-8593
- National Labor College, George Meany Campus, 10000 New Hampshire Ave., Silver Spring, MD 20903-1706, Phone: (800) 367-6724 Fax: (301) 431-5411
- West Virginia University, Safety and Health Extension, 130 Tower Lane, Morgantown, WV 26506-6615, Phone: (800) 626-4748 Fax: (304) 293-5905

Region IV

- Eastern Kentucky University, 521 Lancaster Ave., Room 202, Richmond, KY 40475-3100, Phone: (877) EKU-OSHA Fax: (859) 622-6205
- Georgia Tech Research Institute, 260-14th Street N.W., Atlanta, GA 30332-0837, Phone: (404) 385-3500 Fax: (404) 894-8275
- University of Alabama, 204 Bryant Drive, Tuscaloosa, AL 35487, Phone: (877) 508-7246
- University of South Florida, 2612 Cypress Ridge Blvd, Suite 101, Wesley Chapel, FL 33544, Phone: (800) 852-5362 Fax: (813) 994-1173
- North Carolina State University, 909 Capability Drive, Suite 1600, Raleigh, NC 27606, Phone: (800) 227-0264
- University of Tennessee, 193 Polk Avenue, Suite C, Nashville, TN 37210, Phone: (888) 763-7439 Fax: (615) 532-4937

Region V

- Eastern Michigan University, 103 Boone Hall, Ypsilanti, MI 48197-1699, Phone: (800) 932-8689 Fax: (734) 481-0509
- UAW Health and Safety Dept., 8000 East Jefferson Ave., Detroit, MI 48214-3963, Phone: (800) 932-8689 Fax: (734) 481-0509
- University of Cincinnati, Genome Research Center, 2180 E. Galbraith, 3rd Floor, Rooms 351-378, Cincinnati, OH 45237-1625, Phone: (800) 207-9399 Fax: (513) 558-1756
- Indiana University, 400 E. 7th Street, Room 629, Bloomington, IN 47405, Phone: (866) 563-4820
- University of Wisconsin-Whitewater, 800 West Main Street, Whitewater, WI 53190, Phone: (866) 563-4820
- Ohio Valley Construction Education Foundation, 33 Greenwood Lane, Springboro, OH 45066-3034, Phone: (866) 444-4412 Fax: (937) 704-9394
- Sinclair Community College, 444 W. 3rd St., Dayton, OH 45402-1460, Phone: (866) 444-4412 Fax: (937) 704-9394
- Construction Safety Council, 4100 Madison Street, Hillside, IL 60162-1768, Phone: (800) 552-7744 Fax: (708) 544-2371
- National Safety Council, 1121 Spring Lake Drive, Itasca, IL 60143-3201, Phone: (800) 621-7615 Fax: (630) 285-1613
- Northern Illinois University, 590 Garden Rd., RM 318, DeKalb, IL 60115-2854, Phone: (800) 656-5317 Fax: (815) 753-4203

Region VI

- Southwest Education Center, Texas Engineering Ext. Service, 15515 IH-20 at Lumley, Mesquite, TX 75181-3710, Phone: (800) 723-3811 Fax: (972) 222-2978
- The University of Texas at Arlington, 140 West Mitchell, Arlington, TX 76019-0197, Phone: (866) 906-9190 Fax: (817) 272-2556

Region VII

- Metropolitan Community Colleges, Business & Technology Campus, 1775 Universal Avenue, Kansas City, MO 64120-1313, Phone: (800) 841-7158 Fax: (816) 482-5408
- Kirkwood Community College, 6301 Kirkwood Blvd. SW, Cedar Rapids, IA 52404-5260, Phone: (800) 464-6874 Fax: (319) 398-1250
- National Safety Council, 11620 M Circle, Omaha, NE 68137-2231, Phone: (800) 592-9004 Fax: (402) 896-6331
- Saint Louis University, 3545 Lafayette, Ste. 300, St. Louis, MO 63104-8150, Phone: (800) 332-8833 Fax: (314) 977-8150

Region VIII

- Salt Lake Community College, 391 Chipeta Way, Suite C, Salt Lake City, UT 84108, Phone: (801) 581-4055 Fax: (801) 585-5275
- Uintah Basin Applied Technology College (UBATC), 559 N. 1700 W., Vernal, UT 84078, Phone: (435) 725-7100 Fax: (435) 725-7199
- University of Utah, 391 Chipeta Way, Suite C, Salt Lake City, UT 84108, Phone: (801) 581-4055 Fax: (801) 585-5275
- Rocky Mountain Education Center, Red Rocks Community College, 13300 West Sixth Avenue, Lakewood, CO 80228-1255, Phone: (800) 933-8394 Fax: (303) 980-8339

Region IX

- California State University Dominguez Hills, College of Extended and International Education, 1000 E. Victoria St., Carson, CA 90747, Phone: (888) 4LA-OSHA Phone: (310) 243-2425 Fax: (310) 516-3753
- Chabot-Las Positas Community College District OSHA Training Center, 7600 Dublin Blvd, Suite 102A, Dublin, CA 94568, Phone: (866) 936-OSHA (6742) Fax: (925) 560-9458
- College of Southern Nevada, Division of Workforce and Economic Development, 2409 Las Verdes Street, K1B, Las Vegas, NV 89102-3880, Phone: (877) 651-OSHA (6742) OR (702) 651-4551 Fax: (702) 651-4538
- University of California, San Diego, 404 Camino del Rio South, Suite 102, San Diego, CA 92108, Phone: (800) 358-9206 or (619) 260-3070 Fax: (619) 294-3861

Region X

- University of Washington, 4225 Roosevelt Way NE, #100, Seattle, WA 98105-6099, Phone: (800) 326-7568 Fax: (206) 685-3872

TIME STUDIES, DILO

A time study is one tool that can be used to identify gaps in value-added versus non-value-added work. Once a time study has been completed, cycle times for standard work should be identified and mapped out on a visual board. Time studies are completed to determine cycle time and takt times when identifying waste in a process. Some companies have taken time studies to the next level to identify what issues are facing employees that prevent them from completing job tasks. A "Day-in-the-Life-of" (DILO) study is an all-day study that gives a real-time view of a period of time in the life of management and operations. It analyzes two dimensions that all employees face: (1) concern for tasks (work) and (2) concern for people or employees. It indicates how much time is spent on value-adding and non-value-adding activity and why.

DILO studies can help management do the following:

1. Understand and quantify how supervisors and other employees spend their time
   a. Value add—contributes to the bottom line
   b. Non-value add—does not contribute to the bottom line
2. Determine improvement opportunities in specific target areas
3. Obtain hard data when data needed is not available or there is a concern with data validity
4. Expose management to the employee's real world at a root level to understand their activities
5. Helps determine what we can do about a finding (important insight) during the analyses and to help begin to scope the project
6. Helps understand what the results will really be, and further validate findings

Having other employees help with time studies works well in manufacturing and encourages team building and process knowledge. Professionals can also gain valuable insight into their own workday by completing a time study. It can be as simple as estimating the time to complete daily tasks with the use of Microsoft Outlook to track meetings, appointments, and time.

During a DILO study at a manufacturing facility, it was determined that supervisors spent on average 34% of the workday in meetings regarding the manufacturing process. When categorized by specific tasks, only 8% of the supervisors' time was considered value-added supervising. As data was collected for the study the following categories were identified and established:

- Supervising employees
- Updating payroll/work time data
- Updating information, turnover reports, shortages, etc.
- Meetings
- Continuous improvement projects, lean initiatives
- Scheduling
- Quality
- Lunch and breaks
- Firefighting, chasing issues

DILO studies should be used not to identify standard work for the employee, but rather as a tool to identify processes that need improvement. A process that is unstable, whether in a manufacturing or service industry setting, will identify large chunks of time dedicated to non-value-added work. One way to stabilize a process is with the development of standard work.

STANDARDIZED WORK FOR EHS

What can be considered standardized work for safety? The answer is, again, it depends. Standardized work can be broken down into many different areas, from the hourly line workers to the presidents and CEOs of organizations. Good EHS professionals will identify gaps in their EHS programs and help develop standard work for all employees based on the level of responsibility.

Standard work involving EHS for hourly workers could be as simple as performing forklift or crane inspections on the shop floor. Each process can be different, with different standard work identified that needs to occur from a regulatory standpoint as well as proactive safety awareness point. Some examples of standard work for EHS that can be part of an employee's day include the following:

- Respiratory cleaning and storage
- Storing flammables and combustibles back into storage cabinets
- PPE inspection, including fall protection
- Forklift inspections
- Crane inspections
- Safety device operational checks (e.g., light curtains, interlocks)
- Review of any job safety analysis (JSA) material
- Performing safety audits or inspections as needed or identified

When standardized work is being identified for the hourly employee, EHS must be involved in the process. Once hourly or line employee standard work has been identified, we can start to look at what is called leadership standardized work. Leadership sets the example of standardized work and should also have elements of safety and health identified.

Dr. Dan Petersen is considered one of the most influential people in modern safety and health. Petersen authored numerous books and articles regarding his study of safety and health management and problems associated with EHS management models. Dr. Petersen offered many different models that lend themselves nicely to developing leadership standardized work for EHS. In *Measurement of Safety Performance*, Petersen describes his process, known as SCRAPE or System of Counting and Rating Accident Prevention Effort (Petersen 2005).

The SCRAPE program is a process of measuring what a supervisor does for the EHS program. The process should be modified to fit the business needs of the safety program in what needs to be measured to promote safety and prevent injuries.

The following areas are included in Petersen's model:

- Departmental inspections
- Training and coaching employees

- Accident investigations
- Individual employee contacts
- Safety meetings
- New employee orientation/OTJ training

Petersen also identifies the *menu* method of measuring safety performance. The menu idea is to give supervisors and managers a menu of tasks to select from regarding safety and health. In this model, management will have the ability to select tasks they feel comfortable with and tasks they know will work better with employees. Using menu models also gives management the feeling of control; managers can pick and choose how to measure safety performance for their area or department.

As the scope of the management function broadens, the menu or standardized work will broaden as well. Unfortunately, many organizations fail to sustain good processes and tend to get away from standardized work for safety. British Petroleum (BP) is a prime example of how organizations can fail to sustain safety processes and procedures. Top management at BP equated safety performance to individual injuries such as slips, trips, and falls by individual employees (von Aschwege 2010). Companies with huge process safety management (PSM) programs need to also measure these programs to ensure that performance goals are being met.

The question that needs to be posed to top management is, "What EHS goals have been identified and does top management truly understand the metrics being measured?" Dr. Petersen passed away in 2007, but he left a lasting legacy of how to identify and measure safety performance in any organization using many different approaches and techniques. Clearly ahead of his time in thinking of safety management and measurement, he will be missed.

STANDARDIZED WORK FOR EHS PROFESSIONALS

How does standardized work fit into a chapter regarding managing change, stress, and innovation? Easy—any time EHS professionals can push EHS tasks back to other employees, supervisors, and managers, it lightens the load. It will also force other departments and managers to identify safety metrics within their own department and control these issues. In Chapter 3, we discussed how during 5S activities we create a standard workplace. We also agreed that a standard workplace is safer than a workplace that has not been 5S'ed. The same issue applies with standard work. If we create a standard workplace and identify and track standard work, the process or workplace becomes stable. Everyone within the process understands what needs to be done and by when. The next question to the EHS professional is, "What is your standard work?"

Once standard work has been identified and documented, a standard work board can show and demonstrate what work needs to be completed by when on a daily basis. The board should be visible to all employees in the EHS department as well as other departments in the organization. In Figure 5.2, standard work has been identified and made visual with the use of a dry erase board and magnets. Once standard work has been made visual, it becomes quite clear what needs to be done and by when. In this example, as with many others, weekend days have been eliminated

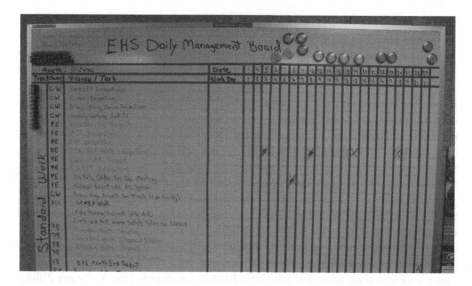

FIGURE 5.2 Example of an EHS department developing standard work board on a white board.

from the board. The theory behind eliminating weekend days from the board is to ensure that time and data management is consistent with the normal work week. If one desires to work weekends, then keep the weekend days on the board. The removal of weekend days also gives a true measurement for time. By removing weekends for tasks to be completed, windows for completion can be reduced by 10 days in some months.

In Figure 5.3, the table is a clearer version of Figure 5.2. Again, as the EHS professional, you will need to identify what standard work needs to be identified and tracked to be successful. As standard work becomes part of the normal work day, the time taken to perform tasks in the past should shorten. As more standard work is defined, the process should become stable, decreasing the effort needed to get tasks completed.

By making EHS standard work visual, it can also give the EHS professional a word that is often frowned upon in the workplace: "No." By placing standard work on a visual board, it lets you make an informed decision on your ability to take on new projects based on time available and work that needs to be completed. It is a lot easier for someone or a department to accept the refusal of added work due to interference with the standard work that needs to be completed. Usually when the "drop everything and do this" phrase is used, it means that another process is unstable. As with any company, there is always one department that has unstable processes on a frequent basis. If this is the case, from a lean perspective you need to try to help stabilize the process. This might mean getting out of the comfort zone of EHS to help other departments. If this is the case, it should be looked at as an opportunity to understand another part of the business or organization.

At the end of the day, people will naturally resist change due to several different thought processes such as fear of the unknown or a fear of losing control. As stated at the beginning of the chapter, the best way to manage change is to create change.

Month: March	Date	1-Mar	2-Mar	3-Mar	4-Mar	7-Mar	8-Mar	9-Mar	10-Mar	11-Mar	14-Mar	15-Mar	16-Mar	17-Mar	18-Mar	21-Mar	22-Mar	23-Mar	24-Mar	25-Mar	28-Mar	29-Mar	30-Mar	31-Mar
Type / Owner	Process/Task \ Workday	1	2	3	4	5	6	7	8	9	10	11	12	13	14	15	16	17	18	19	20	21	22	23
Standard Work																								
CW	Collect Forklift Inspections																							
CW	Collect Crane Inspections																							
PE	Weekly Safety Audits																							
PE	Monthly Air Report																							
CW	AST Inspection																							
CW	RMP Inspection																							
PE	7-Day Haz Waste Inspection																							
PE	IWW Report																							
PE	KARST Inspection																							
PE	EHS Slides for Ops Meeting																							
PE	Month End EHS Report																							
CW	Collect and enter work hours																							
CW	Monthly Env Project Review																							
PE/CW	GEMBA Walk																							
CW	Enter Training Records																							
CW	Create Weekly Safety Talks																							
PE	Haz Waste Sampling																							
PE	Haz Waste Pickup																							
VSR	Pre-promotion resources																							
PE/CW	Supervisor Board Validation																							
PE	Lockout/Tagout Audits																							
PE	Create Safety Newsletter																							
Parking Lot	Update HazCom Training																							
	Update Intranet Site																							
	Battery Recycling Program																							
	Create training bins																							
	Write a book																							
	Review current PPE usage																							
	Revise safety talks																							
	Create cell phone policy																							

FIGURE 5.3 Example of a standard work board in Excel. Development of standard work along with a value stream roll as well as a parking lot area for projects that need to be completed.

If you are not creating change, help manage it by step-change process management. Let everyone know what is going to change and why, review the change, and set an expectation after the change has occurred. Take every chance you have to build relationships in other departments and answer questions regarding EHS.

WORKS CITED

von Aschwege, Tom. *Safety at Work Blog.* May 16, 2010. http://safetyatworkblog.wordpress.com/2010/05/16/bp-safety-culture-and-integrity-management/ (accessed February 23, 2011).

Occupational Safety and Health Administration. *Outreach Training Program Guidelines.* February 2009. http://www.osha.gov/dte/outreach/construction_generalindustry/general_industry.html (accessed March 7, 2011).

Petersen, Dan. "SCRAPE." In *Measurement of Safety Performance*, by Dan Petersen, 59–61. Des Plaines: American Society of Safety Engineers, 2005.

6 Foundations of Individual and Group Behavior in EHS

If you don't know where you are going, you will probably end up somewhere else.

—Laurence J. Peter

As people move through their careers, they become very well versed in different personal and group behaviors, usually with different organizations. The challenge of the EHS professional is to understand and harness different behaviors and philosophies, to "move the needle" in a safety and health program. It's hard to believe that with the exception of a few state-run OSHA and EPA plans, nearly all environmental health and safety standards are the same. Are they the same? Why do different people and companies comply with the same standards in different ways? The answer lies within people's or organizations' cultures and behaviors.

There have been many tests allocated by human resources departments to determine personality types to encourage teams to work better as individuals and as groups. As many organizations look for the "silver bullet" of group dynamics and behavior, personality testing has become standard in larger organizations.

The Myer-Briggs Type Indicator assessment (MBTI) has been used in many Fortune 100 companies, including Apple and Honda, just to name a few. The MBTI identifies and reviews 16 different personality types based on four different dimensions (Robbins and DeCenzo 2005). These dimensions include the following:

- Extraversion versus Introversion (EI)
- Sensing versus Intuition (SN)
- Thinking versus Feeling (TF)
- Judging versus Perceiving (JP)

These dimensions are then broken down and cross-referenced, using different questions to identify personality traits. The end game to personality traits is that some organizations and managers believe that understanding personality traits will help employees and teams of employee work together better to solve problems. Although many companies start out with the best intentions in completing these tests, very rarely are results followed through when personalities have been identified. In most cases, the testing and training for MBTI is for the benefit of the employees to understand their own personalities, as well as the other personalities in the room. These results are rarely used when placing different people on problem-solving teams.

Many other personality tests have been identified and used to identify employees for placement into certain occupations. Not one test has been identified as a catchall to identify personality flaws and proper placement.

WORKING TEAMS AND GROUP BEHAVIOR

Teams that work closely together consistently will have a higher degree of success due to the fact that personality traits are well known between team members. In many emergency services operations, team building occurs on a regular basis with constant work drills, training, and service calls. The question that must be asked by management is why other teams in other industries cannot perform at the same level. Can it be the sense of urgency in emergency services that drives a team for better results? Can it be that, at one time or another, either your life or a coworker's life was at stake during a service call?

I myself at one point in my career was in the fire service. Years later, after leaving the fire service, I was at a team-building and training session for a Fortune 100 company. The group I was in was asked if anyone had ever been part of a high-functioning team that achieved its goal every time. Several people in the training session raised their hands and gave examples of why they thought their team was so successful. When I was asked to give examples of why I thought my team was successful, I stated, "We got called when something was on fire, and we got there and put the fire out. We were successful every time." The trainer went on to elaborate how different personalities play out during service calls just like in business and team dynamics. I stated to her and the rest of the training class that "personalities were there and are always there. The difference for us on the fire scene was that there was work to be done. Someone needed help and that was bigger than any personality or group dynamic that might have been there on that particular call. What ordinary people in emergency services and, to a very large extent, the military do every day is understand that they are a part of something special and much bigger than themselves." A hush fell over the training class after my statement. Although very profound, this is purely my opinion. The difference between high-functioning teams and teams that struggle is the individuals' understanding of how their actions and decisions impact the team and the overall big picture. The key to individual success as well as team success will hinge on how you understand your attributes and use them to your benefit and your team's benefit.

SELECTING AND BUILDING TEAMS

When we talk about lean and kaizen events, we always want to include a cross-functional team in these events. Allowing someone to come from an office environment out to a production or manufacturing area always gives a fresh opinion and idea on a process. Lean isn't just for the manufacturing floor, but for all processes in an organization. Depending on what has been identified as a deliverable out of a kaizen event, teams can be completely cross-functional across an organization or department-specific. A good example of a cross-functional team is a musical band. Each member of the team plays a different instrument, but together, working on a common goal, they create an ensemble.

Another idea of building cross-functional teams is the assignment of strategic planning and objectives. In many cases, policies and programs will reach across department lines and have ramifications throughout an organization. The creation of the cross-functional team to solve a problem helps eliminate any confusion as to what the goal or objective is, ensuring that all responsible parties are in the loop. In many organizations, safety committees and teams are made up of cross-functional disciplines throughout an organization. The need to have a member of a maintenance department on the team can be to ensure that corrective action and work orders are prioritized properly to correct unsafe conditions. The same can be said for an annual review of lockout-tagout procedures with process equipment. In some state OSHA programs, safety teams or committees are required to be in place if a location or organization has more than 10 employees. In Oregon OSHA's *Quick Guide to Safety Committees and Safety Meetings*, exactly how the committee or team should function and what it should accomplish are spelled out by the state program. According to Oregon OSHA (Oregon OSHA):

- Selection
  - If your business has 20 or fewer employees, your committee needs at least two members.
  - If your business has more than 20 employees, your committee needs at least four members.
  - The safety committee must have equal numbers of employer-selected members and employee-elected (or volunteer) members.
  - An employer-selected member can be a manager, supervisor, or any other employee management chooses to serve on the committee.
  - Employees can elect another employee or a supervisor to represent them.
  - If everyone on the committee agrees, there can be a majority of employee-elected members or volunteers.
  - The safety committee can't have a majority of employer-selected members.
- What is required of the Safety Committee—Safety committee members must:
  - Agree on a chairperson.
  - Serve a minimum of 1 year, when possible.
  - Be compensated at their regular pay rates.
  - Be trained in accident and incident investigation principles and know how to apply them.
  - Be trained in hazard identification.
  - Receive safety committee meeting minutes.
  - Represent the major activities of the company.
- Meeting requirements—Monthly or quarterly depending on business type
  - If your employees do mostly office work, meet quarterly.
  - All other employers meet monthly.
  - The safety committee must meet on company time.

- The safety committee doesn't have to meet during a month when you do a quarterly workplace inspection.
- The meeting can be conducted with a conference call, if necessary.
- Records of each meeting should be kept for 3 years, and include:
  - Meeting date
  - Attendees' names
  - Safety and health issues discussed; include hazards involving tools, equipment, the work environment, and work practices
  - Recommendations for correcting hazards and reasonable deadlines for management to respond
  - Name of the person who will follow up on the recommendations
  - All other committee reports, evaluations, and recommendations
- Have procedures for workplace safety and health inspections
  - The safety committee must have procedures for conducting workplace safety and health inspections, including where the inspections are conducted, who conducts the inspections, and how often.
  - Those who do inspections must be trained in hazard identification.
  - Those who do inspections do not have to be safety committee members.
- The Safety Committee needs to accomplish the following tasks:
  - Work with management to establish accident investigation procedures that will identify and correct hazards.
  - Establish a system for employees to report hazards to management and suggest how to correct hazards.
  - Establish a procedure for reviewing inspection reports and making recommendations to management.
  - Evaluate all accident and incident investigations and recommend how to prevent them from happening again.
  - Make safety committee meeting minutes available for all employees to review.
- Evaluate how management holds employees accountable for working safely and recommend ways to strengthen accountability. Examples include evaluating the effectiveness of safety incentives, disciplinary policies, and employee participation in identifying hazards.

Different organizations use different tools to select teams and promote safety. The use of safety committees and teams to create employee buy-in is essential to a successful safety program. Depending on how an organization views committees, safety teams may be a better choice. The word *committee* can have implications of elections and voting processes. In addition, to remove a member of a committee would also require the same (i.e., majority) vote to remove a poor-performing member. Safety teams allow an EHS department some autonomy in selecting and utilizing different employees in different functions. How a safety team functions within the lean environment can have significant impacts on safety, as long as consistency is kept in check.

In 2001, a report from the UK Health and Safety Executive reviewed two very different and distinct views of teams working in lean environments and the impact on health and safety. In one view on teamwork, a reduction in work-related stress occurred through the team effort, enabling greater control over everyday issues that workers faced. The second view offered a much different view that work stress was actually created as workloads increased and created unstable expectations for employees. In the lean perspective, the report indicated that teams were successful and had a positive impact on job autonomy, skills creation, and feedback. The same report indicated that the removal of job tasks during lean events and the reallocation of job tasks to other employees created additional stress (OH Editorial Staff 2001).

There is no question that teams and team building are an essential part of management and EHS processes. Consistency and open lines of communication are the keys to the success of the team. Challenge your teams and reward and praise whenever possible.

GROUPTHINK

Groupthink is "a type of thought within a deeply cohesive in-group whose members try to minimize conflict and reach consensus without critically testing, analyzing, and evaluating ideas. It is a second potential negative consequence of group cohesion" (Wikipedia 2011). Many defining moments in history have been attributed to groupthink issues, including the Challenger Shuttle disaster, attacks on Pearl Harbor, and escalation of U.S. troops in Vietnam during the Johnson administration.

In later reviews of the Challenger Shuttle incident, it was determined by many experts that a consensus agreement was reached to launch Challenger. NASA was under increasing pressure to launch after repeated launch scrubs in the previous days. Engineers were concerned about the temperature, which was abnormally cold for the Kennedy Space Center in Florida. The engineers' concerns were overridden by NASA management as the team or group consensus decision was that the shuttle was safe to launch. Unfortunately, 28 seconds into flight a catastrophic failure occurred in one of the rockets, which caused the vehicle to blow up. Upon further review and interviews in the following months, it was believed by many that groupthink had enveloped the launch team.

In the case of President Lyndon Johnson, the decision to escalate the war in Vietnam led to significant changes in the way U.S. presidents would manage affairs in the future. It is believed by many that Johnson was the victim of groupthink from military advisors and was given bad advice on how to navigate U.S. policy through the war effort. Johnson was repeatedly told that more troops were needed and that the war would turn in the United States' favor if more troops and resources were allocated to the effort. This was in stark contrast to many people outside Johnson's inner group who were analyzing current events.

Identifiable events such as the escalation in Vietnam have led presidents to include what are now called White House advisors or senior counsel members. These people are usually somewhat ordinary people who may have helped the president become elected. Some have been college roommates and lifelong friends. The role that these people play for the president is to be an in-house check and balance to groupthink. You will often hear U.S. presidents talk about the "bubble." The bubble refers to the shield created around an acting president. A vacuum of sorts is created due to the fact that the president can't simply walk into any coffee shop in the country and find out what's really going on. What the bubble has created is a possibility that information will be skewed coming from inside this bubble. Decision making could be compromised if the president does not have someone he or she can trust outside the bubble.

Groupthink can become an issue for EHS programs. Recent events, such as the BP Deepwater Horizon disaster, are currently being reviewed to see if groupthink played a role in the decision to drill when safety measures were not properly in place. Although relatively rare, groupthink still does occur from time to time. Understanding what groupthink is and what it can do will help prevent future issues from occurring.

MOTIVATING AND REWARDING EMPLOYEES

Almost everyone has been a part of some type of team and most likely has been rewarded in some way. As children, we are rewarded in a variety of different ways in different team settings. Team sports, Boy Scouts, Girl Scouts, 4H—you name it and there is some type of reward and motivation factor for the individual and the team.

As we get older, motivation and reward become harder to come by. This is due in part to the fact that as we age, responsibility increases. Behaving responsibly is no longer going above and beyond what was expected of you, but rather part of your daily routine as an adult. The question for all managers is how to motivate employees at work. The reward aspect is the easy part in many companies. Bonus plans as well as merit raises usually play a big part in the overall motivation and reward system with most employees. Cash is king. In today's economy, as this book is being written (2011), many companies have asked employees to compromise on merit increases, bonus payouts, and in some cases fringe benefits. How do employers motivate employees when there is no money to give? In one company, part of the bonus structure had criteria for the amount of recalls a division had in the company. The bonus plan was based on performance for a calendar year. In late January, one of the facilities in the division announced a product recall. This event basically put any type of performance bonus out of reach for all employees, who were only weeks into a new bonus plan. If an incentive or reward plan is used, free vacation days, paraphernalia such as T-shirts, mugs, and key chains, and luncheons can all be forms of a reward or recognition program.

When motivating employees from a safety and health standpoint, many different strategies can be used. There is, however, much confusion about incentives, rewards, and motivating tools and models. There are many programs and items out there on the market and on the Internet for safety incentive programs. The question that always comes up is whether to have a safety incentive program or not. Most experts and safety professionals can agree that safety incentive plans have many pitfalls, including underreporting of injuries, entitlement, and a focus on the prize, rather than building safety into the process.

Questions are also raised as to the meaning of incentive versus reward. What is the incentive for people to come to work every day? What is the incentive to come to work every day on time? The incentive to come to work every day is to keep your job. Many people don't see the incentive to work safely, unless a "significant emotional event" has occurred to either themselves or someone they know. Many companies and managers against safety incentives view working safely as a condition of employment. It is expected that employees work in a safe manner and follow all safety policies and procedures, so why reward people for something they should be doing anyway?

This same philosophy can be used when employee discipline comes into question. If working safely is a condition of employment, why aren't more people terminated for having multiple workplace injuries? If you listen very closely, you can hear HR people gasping for air right now after that statement. The fact of the matter is that many companies continue to employ people with horrible safety records and attitudes. In one facility, there was an employee who had incurred 17 lost-time incidents within a 10-year period. When asked why the employee was still working, the reply from management was, "We can't fire employees for accidents, and he's just unlucky."

However, if the same employee had 17 quality issues in the same time period, chances are the same employee would have been released from his duties. To make a statement that this employee is unlucky is an understatement. Employees with "frequent flyer" miles (meaning that they have been to the clinic or hospital so many times they know people by name) are sometimes the hardest employees to get through to. How do you motivate these employees? The same way you motivate them to come to work every day: put them on a Safety Improvement Plan (SIP).

SAFETY IMPROVEMENT PLANS

The SIP is a program designed to help individuals who have had more incidents and injuries than average, based on the overall number of incidents and injuries in the company. This help will come in the form of involvement to increase awareness and responsibility, thus resulting in improved safety behaviors. SIP is used when an individual employee suffers more than an average number of preventable injuries and incidents than his or her peers. For purposes of SIP, ergonomic or musculoskeletal-related injuries are not counted, unless the employee has previously been advised of preventive actions and has failed to utilize those measures.

The SIP process is not intended to be used as a "punishment" for violation of safety rules and policies, nor solely as a result of recordable injury. SIP is intended as a tool to help the employee who is more accident-prone to examine his or her safety attitude and increase awareness. Ergonomic injuries will not be counted in the SIP plan, unless such ergonomic injury was wholly preventable by the employee (such as failure to use prescribed tools and equipment).

A criterion needs to be developed to determine who would fall into this program. Leadership, EHS, and HR should always work together to determine if the employee should participate in SIP. Criteria can be as simple as stating that any employee who has a rate of two or more accidents or incidents within 6 months or recordable incidents within 12 months needs to be involved in SIP. The SIP program can also be used if a particular supervisor or manager is incurring a large number of incidents or injuries related to EHS programs. This would be considered a Leadership Level SIP. Clear expectations should be made in the SIP for failure to complete the plan with statements such as, "Anyone who is placed in SIP and fails to satisfactorily complete their plan, fails to participate, or continues to display unsafe behaviors is subject to disciplinary processes." Employees should also be allotted time in the normal work-day to complete tasks related to the SIP.

Each department in the organization has a responsibility to the employee in the SIP program.

- EHS
  - Identify employees and supervisors who may need SIP, based on trending.
  - Guide employees to information and access to materials to help in the SIP process.
  - Assist other management members with inspections and audits to improve their performance.
  - Identify employees whose injuries were solely preventable by behaviors (may need referral to Employee Assistance Program [EAP]).
  - Eliminate employees whose injuries are ergonomic in nature and not wholly preventable by employee behaviors.
- Human Resources
  - Consult with EHS, plant leaders, and employees to help establish reasonable and meaningful SIP goals.
- Management
  - Assist supervisors in developing and encouraging safe behaviors and improvement efforts.
  - Consult EHS where concerns on risk exist.
- Supervisors
  - Help to guide peers and employees to improve safety in their areas.
  - Communicate SIP progress to involved employees.
  - Communicate personal SIP progress to leader.
- Employees
  - Take actions to correct unsafe situations and improve safety.
  - Develop and implement SIP where required.
  - Support coworkers in their SIP efforts.
- Communicate progress to leaders.

Obviously, for an employee to get to this point there has been some discussion associated with either unsafe acts and/or conditions. Care should be taken when discussing the issue of placing an employee into an SIP program. Meeting with the potential participant's supervisor and human resources should occur to discuss situations where an employee may have other reasons for not performing safely and to determine if participation in SIP will help an employee to succeed. Once it has been agreed that the participant can benefit from the SIP program, a meeting should be held to review the following:

- Why has the employee been selected for the program?
- What is the Safety Improvement Plan (SIP) all about?
- What are the expectations of the plan?
- When does the plan begin?
- Where does it end?

To give the participant some sense of control, a menu of safety-related tasks can be developed for the participant to pick from for completion of the SIP. Some of these tasks can include the following:

- Conduct four department safety inspections.
- Participate in department safety training.
- Implement two or more safety improvements for their department.
- Give a safety talk to your department.
- Complete a work site analysis for a specific work area.
- Complete a job safety analysis for a specific hazardous job.
- Identify four unsafe conditions or acts happening in your area, and determine ways to prevent them.
- Attend safety committee meetings for a given period of time.
- Assist in a hazard assessment.

At no time should any items on the SIP menu be identified or viewed as punishment for workplace injuries. The SIP should be used as a vehicle to teach and learn from all perspectives. The SIP should be drawn up as a contract between the participant and the organization to eliminate any guessing as to what the expectation is from the participant and everyone else involved in the process. Figure 6.1 shows an example of what a SIP contract or agreement can look like for employees.

The same process can be used with a supervisor, but with a different menu selection. The expectation of a supervisor or manager placed into an SIP program is and should be significantly greater. Acceptable action items and deliverables that can be part of the supervisor/manager SIP menu can include the following:

- Attend a safety seminar or class related to the department's identified issues, and report on findings.
- Implement four or more safety improvements for their department.
- Prepare three toolbox talks for sharing with the entire company.
- Complete four worksite analyses.
- Complete four job safety analyses.
- Increase behavior observations.
- Conduct behavior observations with the plant manager and/or safety department.
- Participate in an executive safety meeting.
- Conduct a department hazard assessment.

In Figure 6.2, an example of a supervisor/manager agreement is identified.

The process of the SIP should always be the same in that there must be closure to the process. The participant, EHS, HR, and the participant's leader will agree on the final objective or goal for the participant. The employee will collect documentation on his or her participation and present it to HR and the supervisor for review. Upon successful completion of the course of action, a review by the EHS department will be conducted to check improvement. Consequences of noncompliance should also be addressed to ensure that anyone who chooses to stop participating in his or

Safety Improvement Plan Agreement Non-Supervisor Program

I, _____ understand that I have been identified as an "at risk" employee under

Safety Improvement Plan. **Under the SIP Program, Section 3.2** *"Any employee who has a rate of 2 or*

more accidents or incidents within six months or recordable incidents within 12 months will be identified as

needing to be involved in SIP".

I understand that I have incurred the following incidents that have identified me as an "at risk" employee.

#	Date	Injury Type	Description of Injury
		☐First Aid ☐Recordable ☐Lost Time ☐Property Damage	
		☐First Aid ☐Recordable ☐Lost Time ☐Property Damage	
		☐First Aid ☐Recordable ☐Lost Time ☐Property Damage	
		☐First Aid ☐Recordable ☐Lost Time ☐Property Damage	

I, _____ have chosen to complete the following actions to fulfill my obligation

to the SIP Program.

☐Conduct 4 department safety inspections.
☐Participate in department safety training.
☐Implement 2 or more safety improvements for their department.
☐Give a safety talk to your department.
☐Complete a work site analysis for a specific work area.
☐Complete a job safety analysis for a specific hazardous job.
☐Identify 4 unsafe conditions or acts happening in your area, and determine ways to prevent them.
☐Attend a safety committee meeting.
☐Assist in a hazard assessment.

I, _____ understand that If at any time I choose to stop participation in, or
activity related to this SIP, I will be subject to additional action to include disciplinary action as deemed by
Human Resources.

SIP Participant: _____ EHS Representative: _____

Supervisor: _____ HR Representative: _____

FIGURE 6.1　Safety Improvement Program contract or agreement for employees.

her SIP or activity related to the SIP will be subject to additional action to include disciplinary action as deemed necessary. Figure 6.3 demonstrates the simple process flow of the SIP program.

In lean enterprise, motivating and rewarding employees is built into the culture. As processes become leaner, the time and effort needed to perform the same task

Safety Improvement Plan Agreement, Supervisor Program

I, _____ understand that I have been identified as an "at risk" Supervisor under

the Safety Improvement Plan. **Under Section 3.3 of the Safety Improvement Plan,** *"A supervisor whose*

team has a higher than plant or company average of incidents (recordable or preventable) will participate

in a leader level SIP involving mentoring from the safety office".

I understand that I have incurred the following incidents that have identified me as an "at risk" Supervisor.

#	Date	Injury Type	Description of Injury
		☐First Aid ☐Recordable ☐Lost Time ☐Property Damage	
		☐First Aid ☐Recordable ☐Lost Time ☐Property Damage	
		☐First Aid ☐Recordable ☐Lost Time ☐Property Damage	
		☐First Aid ☐Recordable ☐Lost Time ☐Property Damage	
		☐First Aid ☐Recordable ☐Lost Time ☐Property Damage	
		☐First Aid ☐Recordable ☐Lost Time ☐Property Damage	

I, _____ have chosen to complete the following actions to fulfill my obligation

to the SIP Program.

☐Attend a safety seminar or class related to the department's identified issues, and report out on findings.
☐Implement 4 or more safety improvements for their department.
☐Prepare 3 toolbox talks for sharing with entire company.
☐Complete 4 worksite analyses.
☐Complete 4 job safety analyses.
☐Increase STOP observations.
☐Conduct STOP observations with the plant manager and or safety department.
☐Participate in Executive Safety meeting
☐Conduct a department hazard assessment.

I, _____ understand that If at any time I choose to stop participation in, or
activity related to this SIP, I will be subject to additional action to include disciplinary action as deemed by
Human Resources.

SIP Participant: _____ EHS Representative: _____

Plant Director: _____ HR Representative: _____

FIGURE 6.2 Example of a supervisor/manager agreement for a Safety Improvement Plan.

should be reduced. The reward or incentive at that point is to make your job easier, ergo reducing the amount of stress in your job. Praise kaizen teams for the time spent improving the process and demonstrate your support by going to report outs and asking value-added questions.

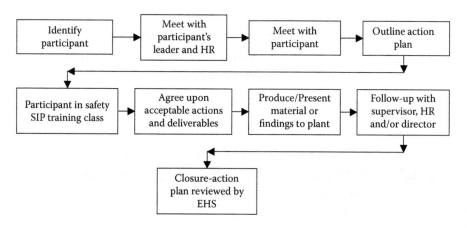

FIGURE 6.3   A simple process flow of the SIP program.

WORKS CITED

OH Editorial Staff. *New Report Examines Teamwork and Stress.* December 12, 2001. http://ehstoday.com/news/ehs_imp_35004/index.html (accessed January 21, 2011).

Oregon OSHA. "Oregon OSHA's Quick Guide to Safety Committees and Safety Meetings." *orosha.org.* http://www.orosha.org/pdf/pubs/0989.pdf (accessed March 15, 2011).

Robbins, Stephen P., and David A. DeCenzo. "Basic Organization Designs." In *Fundamentals of Management*, 163. Upper Saddle River, NJ: Pearson Prentice Hall, 2005.

Wikipedia. *Groupthink.* April 9, 2011. http://en.wikipedia.org/wiki/Groupthink (accessed April 12, 2011).

7 Leadership

People ask the difference between a leader and a boss. . . . The leader works in the open, and the boss in covert. The leader leads, and the boss drives.

—Theodore Roosevelt

LEADERS VERSUS MANAGERS

Reading the quote above by Theodore Roosevelt makes sense to many of us. Throughout my career, I can think of several people who I thought were leaders and not just managers. If you think about the differences between leading and managing, one can be viewed as the bare minimum effort needed to get through the day— managing. While leading may require extra effort, understanding, and time, the difference between the two can mean success or failure of a team or objective.

I am often amazed at the U.S. military and how, on occasion, ordinary soldiers, marines, seamen, and airmen do extraordinary things. The mind-set of someone in the armed services is that of someone being exposed to leadership on a constant basis. I have always admired and been in awe of people selected for the Medal of Honor. I take the time to read the official narrative on what the recipient did to achieve the highest award possible in military service. Always given for valor above and beyond the call of duty, medals are, sadly, usually awarded posthumously. The narratives reveal that, at some point, the recipients took leadership matters into their own hands and made decisions for a favorable outcome of the group or team. It is for these reasons that many companies and industries prefer employees with military experience. Military training and leadership always teach not to give up and to readjust, adapt, and move on.

Harold Gregory "Hal" Moore, Jr. is a retired lieutenant general in the U.S. Army. General Moore in 1992 wrote a book called *We Were Soldiers Once . . . and Young*, which was Moore's account of commanding the First Battalion, Seventh Cavalry at the Battle of Ia Drang, Vietnam. The Battle of Ia Drang was the first major engagement of the Vietnam War in which U.S. forces and the People's Army of Vietnam (PAV) fought against each other. The book was later immortalized in the movie *We Were Soldiers* in 2002. During the battle under General Moore's leadership, the engagement included the following:

- 450 U.S. soldiers versus 2,000 PAV regulars
- An entire platoon cut off from the rest of the battalion
- One helicopter crash landing
- A friendly fire incident that took American lives
- Numerous ambushes

The leadership principles of General Moore during the battle were the same principles that he instilled in himself and his troops before deployment. In an interview with General Moore by Armchair General, the question was asked how to generalize his leadership style. General Moore stated,

> My leadership philosophy, which I employed in the military and also for years in the civilian sector, can be summarized as "power down." I pushed the authority to make decisions down to company commanders and told them to push the power down to their squad leaders and the individual Soldiers in the ranks. I told my people, "If you feel you are qualified to make a decision or to take action, do it. Otherwise, move it up a notch for a decision." (Moore 2007)

Moore is also credited with creating what is called the Four Principles of Leadership in Battle. These principles include one of the most famous quotes from General Moore, "Three strikes and you're *not* out." Moore relates that three strikes and you're out only works in baseball. There is always something the leader can do to change the outcome of a situation or problem. The time when everything is going according to plan is the time when complacence sets in, threatening the plan. Moore also agrees that leaders should act on instinct and believes that it is easier to ask forgiveness than to get permission.

I am lucky enough to have a family member in the military. I asked him a few questions when I was writing this book, and he shared some insight into what leadership values were instilled in him. When I asked him the difference between military leadership and corporate leadership he told me,

> Corporate leadership usually ends when the employee leaves for the day. Military leadership revolves around the principle that we must be accountable for our actions, and that of our subordinates, 24 hours a day. If a subordinate gets in trouble on the weekend, the subordinates' chain of command is expected to get involved, and must take time out of their weekend to resolve the problem. The leaders are questioned on what they did to prevent the issue before it became a problem (i.e., safety brief, counseling) and what their course of action to correct the problem is. This invests the leaders in their subordinates' professional and personal lives, while placing a greater emphasis on ensuring subordinates are capable of preventing/resolving issues before they become larger problems. Constant accountability and responsibility of all subordinates' actions is probably the most defining characteristic of leadership within the military.

His statement validates the fact that subordinates are investments to the leadership in the military. Does your company look at employees as an investment?

Thank you, Lieutenant. I appreciate your service to our great country and the freedom you provide as well as all others who have served at one time or another.

LEADERSHIP TRAITS AND TRUST

What makes a good leader? There can be many different answers to this question. The difference between leading and managing can be strikingly different. There have been many different studies on leadership, with five different traits identified as leadership qualities.

People want to work with and be led by an honest leader. Honesty, as well as ethics, played a big role in the latest economic downturn. The auto industry appeared on Capitol Hill asking Congress for a loan to float General Motors, Chrysler, and Ford Motor Company. In statements to Congress, all three companies blamed the increased cost of raw materials and a slow economy. Upon further review, Ford elected not to take federal money to stabilize their business. Both GM and Chrysler received a combined total of $17.4 billion in loans (*Wall Street Journal* 2008). Many customers also felt that Ford was extremely honest, and demonstrated that when sales of Ford vehicles started to rise in late 2009. By 2010, sales of Ford vehicles were up by 33% and even posted a 1% gain in market share, as sales for both GM and Chrysler fell (Kell 2010).

Honesty in leadership plays an especially big role when mistakes have been made. During the BP Deepwater Horizon incident in 2010, then CEO Tony Hayward created a dishonest atmosphere, using poor word choices during different interviews. Before the Deepwater Horizon incident, BP had integrity issues due to past catastrophic incidents such as the Texas City explosion in 2005 that killed 15 workers. At the end of the day, people want to work for an honest company and instill honesty as a value and model for all employees to follow.

Another trait that is important to being a good leader is to be a visionary. Forward thinkers are always looking at how to get to the next level, how to get to the ideal business state. Forward thinking is the entire basis of lean manufacturing and enterprise. Lean enterprise reduces variation, eliminates waste, and improves the process. By completing all of these tasks, companies now have time to focus on the future instead of dealing with the present.

When Henry Ford decided to go into the automobile business, there were no paved roads, fuel stations, or repair shops. Ford wrote,

> The Edison Company offered me the general superintendence of the company but only on condition that I would give up my gas engine and devote myself to something really useful. I had to choose between my job and my automobile. I chose the automobile, or rather I gave up the job—there was really nothing in the way of a choice. For already I knew that the car was bound to be a success. I quit my job on August 15, 1899, and went into the automobile business. (Ford and Crowther 1923)

His vision of building a car and building it successfully with no infrastructure in place clearly required forward thinking.

When leaders are identified as not being very visionary or forward thinking, there is a distinct possibility that the leader may be afraid or unwilling to share their vision with others. If someone came to you 15 years ago and told you, "I am going to start a company that will produce a new type of phone. I want it to be able to take pictures, double as a video camera, send text messages, intercept e-mail, and play movies," what would you have thought? There is also a tendency for leaders not to share visions in the event that a goal or vision has not or will not be reached. In some cases, leaders have been seen as somewhat "flaky" when a vision is broadcast and then not achieved. Leonardo da Vinci is considered one of the greatest visionaries of all time. However, if you looked at his work during the time he was alive, your impression would have been quite different.

Employees also want a leader to be competent. By nature, employees assume you are competent if you are placed into a leadership position. However, the jury is still out until a few successful demonstrations of competency are under your belt. Leaders must demonstrate competence every chance they get. There is a fine line to this trait; too much competence demonstration can lead to thoughts of arrogance by others. Demonstrating competence also doesn't mean that leaders have to inject perspective into every meeting or conversation. Abraham Lincoln once said, "It is better to remain silent and be thought a fool than to open one's mouth and remove all doubt."

Intelligence is different from competence and is another important trait that employees identify with leaders. The development of intelligence is always ongoing, in both formal and informal settings. Formal settings such as college give people the intelligence and understanding of different subjects. Informal settings such as the workplace also offer the opportunity to teach and learn new knowledge. When I was a safety engineer at Ford Motor Company, my divisional boss and I were walking through one of the assembly plants. He turned to me and said, "I can't teach you anything about safety and health that you don't already know. What I can teach you is how to get things done at Ford." Obviously, this conversation caught my attention as I believed I was going to get a magic answer to solve my issues. What he taught me was how to navigate in the plant for safety and health issues, how to approach different departments and sell safety to other managers. Intelligence will also be demonstrated in how you handle yourself regarding personal behavior and attitude. I have met a lot of really intelligent people who have had horrible attitudes. Everyone knows someone like this. Do you think of these people as good leaders? Do you feel that if they just had a better attitude, they would be great to work for?

Inspiration can tie all of these traits together. People, in general, want to be inspired. The writing of this book was inspired by my own passion for safety and telling my stories. Storytelling is inspiration's biggest tool. It allows you to use visions and paint a picture of what you are trying to say or accomplish. Stories that can connect to employees and people on an emotional level will leave lasting impressions. If you have ever gone to hear a keynote speaker, usually there will be some storytelling to draw the audience in. The more emotional the story, the farther people will go to make the connection on a personal level.

Inspiration and passion from a leader can motivate a workforce if the workforce believes the leader is honest, visionary, competent, and intelligent (Leadership 501 2010).

LEADERSHIP STANDARDIZED WORK

When we create standardized work for employees, we create standard work for all employees, including leaders. Lean is all about change, and leaders must be open to change and create it. How will management support the lean culture? Leaders are people too, and we already established that the human condition is to resist change. This is why it is important to understand what role leadership will have to support lean change. What should be in the leadership standardized work? The answer is, once again, it depends. Mostly, it depends on where the organization is in its lean journey. If a culture

Standard EHS Work	Daily	Weekly	Monthly	Annual
Safety Operating System, SOS Board Review	X			
GEMBA Safety Walk		X		
Lean Steering Committee Meeting		X		
Safety and Health Assessment Review Process, SHARP Element Meeting			X	
Safety Process Review Board, SPRB			X	
SHARP Assessment				X

FIGURE 7.1 Example of leadership standardized work for safety.

has already been built and sustainment is an issue, leadership standardized work will look different from that of an organization just starting to build a lean culture.

At one facility, the leadership standardized work included the following identified in Figure 7.1. Although in Figure 7.1 the leadership standardized work for EHS may appear to lack substance, each piece or element of work was substantial.

SAFETY OPERATING SYSTEM (SOS)—DAILY

- Review of medical compliance issues, audiometric testing, PFTs, respirator fit testing needed
- Review of manager overdue incident investigation reports
- Review of supervisor overdue incident investigation reports
- Review first aid/near miss rate (rolling 12-month)
- Review DART rate (rolling 12-month)
- Review ergonomic fixes and reviews for facility top 25 jobs

GEMBA WALK FOR SAFETY—EVERY MONDAY

The leadership GEMBA walk occurred every day, Monday through Friday from 11 a.m. to 12 p.m. Each day represented a different function of the manufacturing process. Monday was deemed as the day the leadership team would review safety. There were nine critical questions that were asked regarding safety in the area the GEMBA walk occurred. Each question had significance to the overall safety program and culture that was being reinforced. Figure 7.2 demonstrates the specific targeted areas that were covered during the GEMBA walk for safety.

Other areas that were included in the EHS leadership standard work were participating in the annual EHS audit process. Each leader was assigned as a champion of a different element covered in the audit. The champion for a specific audit formed a cross-functional team to ensure that all departments were working toward sustaining EHS programs, practices, and policies within the facility. Safety process review board (SPRB) meetings were also held once a month. These meetings were overall process checks of the safety program to make sure everyone was on the same page and all countermeasures were being addressed.

Leadership standardized work needs to be driven from the top down. Leaders need to be held accountable just like all other employees. Leaders lead by example. In one facility I worked at, we had a plant manager who came in and jump-started

Leadership Standardized Work for EHS GEMBA Walks

#	Expectations	Description	Definitions for basis of Question	Questions to Ask on GEMBA
1	Are daily safety startup checks performed and satisfactory?	• Work station checks to ensure safe operations are maintained. • Should be conducted by the Supervisor or Group Leader, GL. • Checks must be reviewed and signed by the Supervisor.	• Startup safety checks are completed within the first 2 hours of the shift to ensure safe operations are maintained. • These checklists may be conducted by supervision or a member of the workgroup but must be reviewed by a member of management when completed by workgroup members. • Supervisor and or GL should notify the appropriate maintenance personnel and management for immediate correction. Employees and or Supervisor needs to follow-up for completion status.	1. Who performs safety checklists? 2. How do you insure they are done within the first two hours of the shift? 3. Who do you communicate non-conformance to? 4. How do you communicate and handle non-conformance? 5. Where are the action plans/results of improvements?
2	Are JSAs in place, maintained, and validated against the task performed, including feedback to employee when discrepancies are found?	• A document which describes safe practices to perform the identified task • Defines tool requirements and their safe use • Should be posted or accessible to all employees in the area.	• A production JSA (Job Safety Analysis) is intended as an instructional aid to assist production employees in understanding the sequence of steps so that they may perform their jobs safely. • JSA's may be posted at the workstation or provided to employees in a central location with uninhibited access. • Periodically, JSA's should be used to validate proper safe work procedures by comparing them to actual employee practices at the workstation. • JSA's should be reviewed by employees and updated at least annually. • JSA's should be utilized as a training toll for transferring and new employees to the workstation.	1. What is the process if non-compliance is found? 2. Are all JSA in your area signed off per corporate standards? 3. Do you train new operators using JSAs? 4. Is there evidence that JSA is being used as a training tool? 5. Verify by looking at the JSA to see if a training sign-in sheet is present.
3	Are incident records and near miss data up to date including	• Supervisor MUST document all injuries / accidents / near-misses- • Should utilize visual	• Incident data should be reviewed by workgroups and posted in their respective workgroup areas on a regular basis including strategies to address improvement	1. How do you acquire OHSIM reports? 2. Do you review incidents and

Leadership Standardized Work for EHS GEMBA Walks

#	Expectations	Description	Definitions for basis of Question	Questions to Ask on GEMBA
	corrective actions?	charts to track injury data trends within the area	opportunities.	near misses in the morning startup meeting? 3. Are their time bound accountable action plans to close issues?

4	Are 5S process standards followed (i.e., regular housekeeping inspections and "cleaning is inspection")?	Should follow the standard 5S Check sheet process to ensure compliance
		
		Should post cleaning schedules and expectations in the area
		
		Should look for ongoing process stability
		
		Follow the 'everything has a place and everything is in its place' disciplined part of the employees' daily work to ensure their work place is organized and maintained.
			• Production workgroups should be able to demonstrate how the 5S process (Sort, Stabilize, Shine, Standardize and Sustain) is being applied within their respective departments to support safety and housekeeping excellence.
			• Examples could include
			  ○ **Sort:** Safety hazard control through proper stacking, storage and clear aisles;
			  ○ **Stabilize:** a place for everything and everything in place (i.e. Materials and Equipment stored and labeled properly, Red Tag process followed accordingly).
			  ○ **Shine:** maintain cleanliness,
			  ○ **Standardize:** Visual Factory Standards are well defined, specific, accessible by Work Groups and adhered to, including HAZCOM labels, Lockout Placards, etc, and
			  ○ **Sustain:** All 5S procedures are followed regularly (Check sheets, Cleaning activities, 5S Evaluations, Adherence to Standards, and management support and follow-up)
				1. Are visual factory standards available?
				
				2. Where are they located?
				
				3. Do you have the most recent edition?
				
				4. When/who performs the actions to keep the area clean and safe?
				
				5. Where is the 5S Checksheet?
				
				6. What are your 5S standards?

5	Are weekly Safety Audit Checklists conducted and satisfactory?	Should be documented and audits kept on file	• The department specific safety audit process should be implemented. • These reviews should be conducted regularly (minimum once per week), documented and retained on file for a period of 8 weeks. • Audit findings should be posted in the work area including time bound corrective action plans (who does what and by when) to ensure issues are assigned to appropriate personnel and resolved in a timely manner.	1. Who performs the Weekly Safety Checklists? 2. Are the results of the Weekly Safety Checklist shared with the employees?

FIGURE 7.2 Example of GEMBA walk questions for leaders regarding EHS.

Leadership Standardized Work for EHS GEMBA Walks

#	Expectations	Description	Definitions for basis of Question	Questions to Ask on GEMBA
6	Is there an effective system to track and ensure health & safety issues are completed on a timely basis?	Action plans should be posted in the area. Safety should be reviewed and closed within 24 hours when possible.	• Each production area should ensure that an effective process is in place to facilitate the communication of health and safety concerns by employees and/or workgroups to their respective supervision. • Employee and/or workgroup issues should be posted in the work area including time bound corrective action plans (who does what and by when) to ensure issues are assigned to appropriate personnel and resolved in a timely manner.	1. How do you receive Safety Items? 2. Are Safety issues brought to the attention of the Supervisor? 3. Review the Safety Issues Matrix 4. What is the Mean Time to Close?
7	Do work group members have input into ergonomic improvements?	Informal process to ensure group members has the ability to address ergonomic issues. Should have a mechanism that allows group member input into Ergonomic fixes.	• The organization should ensure that production based workgroup members are included in the development and implementation of proactive and reactive health, safety and ergonomic improvement opportunities. • Opportunities for employee input include documentation in areas such as, but not limited to 8-D problem solving teams, local change management process, local ergonomics committees, etc.	1. What is the process to identify/communicate ergo issues? 2. Are ergonomic devices and tools being used properly? 3. Is the use of the ergonomic tool or device called out in the JSA?
8	Is performance tracked on key safety metrics?	MUST be identified as part of policy deployment	• At a minimum, each production workgroup should review, post and monitor their respective injuries on a regular basis within their respective workgroups and areas.	1. What is your Safety Metric? 2. Where do you get the data? 3. Is this something you can impact? 4. If you make an improvement, which plant metric will it impact? 5. What is your target?

Leadership Standardized Work for EHS GEMBA Walks

#	Expectations	Description	Definitions for basis of Question	Questions to Ask on GEMBA
9	Are work group activities resulting in improvements made on key safety metrics?	▪ Performance data is improving over time (generally 3 months of improvements required), with direct correlation to work group activities identified and tracked as part of action planning. ▪ Must be linked to policy deployment.	▪ Workgroup's play a key role in driving health and safety and continuous improvement due to their knowledge and understanding of the operation and related health and safety issues. ▪ In that regard, the workgroup should be able to demonstrate WG activity on specific issues in their area that have resulted in improving trends (for trends reflecting three to more months of consecutive improvement).	1. How long have you been tracking this metric? 2. What are your plans or actions to make it better? 3. Have you seen improvement? 4. When do you get time to problem solve? 5. Are the operators in your area involved? 6. Have you needed outside assistance? 7. How do you communicate that you need help?

FIGURE 7.2 (Continued) Example of GEMBA walk questions for leaders regarding EHS.

the stalled lean program. When leadership participation was lacking on the GEMBA walk one day, he called all managers into the conference room. He talked about how important lean was and how the GEMBA walk was a significant part of leadership's commitment to the lean process. He then said, "Starting today, there will be no more meetings scheduled in this facility from 11 a.m. to 12 p.m. This time is set aside by me for the leadership of this facility to get on the floor. I expect you to be on the floor one hour a day helping sustain our lean processes." Everyone left the meeting and went about their business. About 2 days later at 11 a.m., the plant manager was walking past an office and saw two managers sitting. He went in and asked what was going on. Both managers stated that they were working on a project together. Soon after a "significant emotional event" occurred, reinforcing the need and message that leadership was expected to be visible on the shop floor, driving lean.

LEADERSHIP IN EHS

What makes a good EHS manager, director, or leader? There are all kinds of bench-marking tools on different safety metrics and models for programs and policies. There is a good chance that you already know how to write a safety program or, at the very least, how to get one. Here are some basic concepts and tools that all EHS professionals should remember. These concepts, when used properly, will form and shape a good EHS leader.

SELL, SELL, AND SELL!!!

Selling safety is a huge part of any program. The EHS professional must be able to sway and sell the customer in why they should implement new processes or comply with policy. EHS professionals should not beat other managers over the head with standards. At the same time, when something does go wrong (and it will), resist the temptation to say, "I told you so." Working with management when things do go wrong will build team trust and leadership. EHS professionals need to be visionary and create programs that everyone can understand and comply with.

KNOW YOUR PROCESS AND FLOW

Safety is the same whether you're building cars or cookies. With the exception of some different standards in state-run OSH programs, all OSHA standards are the same. It's applying the standard to a specific process that challenges EHS profession-als. You need to spend time on the floor asking questions and learning the processes of your location or facility. This knowledge gives you the credibility to talk to opera-tors, managers, and engineers regarding safety issues and improvements. If possible, try to do the job task in question yourself. Someone can tell you that it hurts to pick up raw material all day, but until you do it yourself, you will never really know. At the same time, you will also build relationships and trust with employees working on the floor.

I can remember at one facility in my career, we had a new plant manager start one week. On Friday, he wanted to have a meeting with all employees to introduce himself and his vision of where we needed to go. The past plant manager usually held two meetings: one for first shift and one for second shift. At the time, I was in charge of security as well as safety and environmental for the facility. About an hour before the meeting was to take place, I learned that the new plant manager had decided to hold one meeting for everyone at the end of the first shift. In his mind, holding two different meetings was a waste of time and people would not come to the second meeting once information got out regarding the first meeting.

I went to his office after he was on the job for one week. I explained to him that I wished he had consulted me before scheduling the meeting. His response was, "Why in the world would I consult with safety to hold a meeting with all of my employees?" I asked him to come with me to the front door of the facility in the lobby. As we opened the front door, I turned to him and said, "Because I am one of the few people here who can tell you our parking lot can only handle one shift at a time." Looking out the front door, we could see a line of cars trying to get into our parking lot. The plant was built on an elevation, so we could also see cars stretched through the industrial park, across the bridge over the highway, and finally backed up on the highway. The state police showed up and asked why they weren't notified of the meeting. The plant manager looked at me, and I didn't say a word to him. I went over the talk to the troopers and explained it was a communication issue and we would appreciate help with traffic once the meeting was over.

The parking lot looked like a scene from Woodstock. We had cars blocking other cars because employees were late and couldn't find a parking space. Fearing they would get in trouble for clocking in late, vehicles were left in the middle of some lots. Several vehicles ended up in culverts and drainage ditches after employees attempted to park on wet grass.

I never told the plant manager, "Listen to me" or "I told you so." We moved past that day and were able to laugh about it years later. For the rest of my tenure at the facility, I was always included in the decision-making process. The moral of this story is if you know your processes and how they work, you bring value to the organization and leadership to EHS.

Consistency: Do What You Say You Are Doing

To build a strong foundation for the EHS program, consistency is the key and must be practiced every day. Whenever I go to different companies for consulting or benchmarking, I always ask to see written programs. Though I understand that not every organization is perfect, 90% of the time I find discrepancies in the written program as to what is occurring on the shop floor. The whole reason to write safety programs is to make a program understandable to non-safety people and create a program better than the standard.

The fastest way to get in trouble during a compliance visit is to say you're doing something on paper when you're not doing it in reality. If there is part of the program that no one is following and can be omitted from the safety program, omit it. Lean is all about documenting real life. Safety programs should also document real life.

Remember Your Audience

Always remember who you're talking to in both content and meaning. EHS is usually the one department that will deal with issues from the president to the janitor. How you communicate with people will leave a lasting impression. Presidents and CEOs are usually looking at the company or organization at a high level. Shop floor employees and supervisors are dealing with the here and now. EHS professionals need to calibrate communication skills to different levels of meaning and understanding to be successful. If possible, communicate face-to-face rather than through e-mail.

Never Stop Learning

Become the perpetual student. If you are a part of a company that is embarking on lean enterprise or is already engaged in lean, get involved. Read books on lean manufacturing, do research, and improve your own process. Get on a cross-functional kaizen team and learn another part of the business. Participate in DILO studies and get to know what supervisors go through on the front lines in production. Formal learning gives you the foundation; informal learning gives you the tools.

WORKS CITED

Ford, Henry, and Samuel Crowther. *My Life and Work.* Garden City, NY: Country Life Press, 1923.

Kell, John. *Wall Street Journal.* January 5, 2010. http://www.aipnews.com/talk/forums/thread-view.asp?tid=11456&posts=1 (accessed February 3, 2011).

Leadership 501. *Leadership Traits—The Five Most Important Leadership Qualities.* 2010. http://www.leadership501.com/five-most-important-leadership-traits/27/ (accessed March 20, 2011).

Lt. Gen. Harold G. Moore, U.S. Army (Retired). *Battlefield Leadership.* http://www.lzxray.com/battle.htm (accessed February 21, 2011).

Moore, General Hal, interview by Armchair General. *10 Questions for General Hal Moore* (September 21, 2007).

Wall Street Journal. Ford Absent from Auto Bailout but May Be Big Winner. December 19, 2008. http://www.foxnews.com/story/0,2933,470525,00.html (accessed January 3, 2011).

Glossary

Andon: Japanese term for *lantern*. In industry it means a signal light. It is designed as a visual aid to bring attention and action in response to the status of process. Color coding will define if a condition is normal or abnormal.

Autonomation (Jidoka): Provides machines and operators with the ability to detect when an abnormal condition has occurred and immediately stop work.

Available Time: Total shift time (in hours) less any breaks, which equals the actual available production time.

Bottleneck: Any activity or process that restricts flow or limits capacity.

Brainstorming: Suggesting (especially among groups of people) a large number of solutions or ideas and combining and developing them until an optimum solution is found.

Cell, Work Cell: An optimal layout of machines and people for a dedicated product or family of products.

Cellular Manufacturing: A process of manufacturing where the work area layout is in a process sequence. Workers continue to follow the work sequence and a "water spider" presents the materials to them from outside the cell, not interrupting the process.

Constraint: A workstation or a process that limits the output of the entire system.

Continuous Improvement (Kaizen): A philosophy of implementing small improvements every day. Small improvements over time will eventually have a large impact on a business.

Continuous Flow: The concept of producing a single piece in a continuous process to eliminate waste, improve quality, and reduce lead time.

Cycle Time: The fixed time it takes to do one repetition of any particular task. Cycle time can be separated into three categories: manual cycle time, machine cycle time, and auto cycle time.

Eight Types of Waste: Overproduction, excessive inventory, transportation, unnecessary motion, waiting, overprocessing, defects, and underutilized people.

Ergonomics: The study of the body's motions when performing a task. Ergonomics has a direct impact of the effectiveness of the workers. It includes their health and safety in the workplace.

5S Program: Refers to a system that includes five activities for identifying and eliminating waste to help create a standard workplace. (1) Sort—separate needed items to complete job task from unnecessary items. (2) Set in order—identify and mark where needed items will be stored. (3) Shine—clean work area and equipment. (4) Standardize—create standard workstations when possible to eliminate variation. (5) Sustain—make sort, set in order, shine, and standardize part of the normal work routine.

5-Why: 5-Why is a problem-solving process that involves asking why to the problem statement. It usually requires five whys to identify the root cause of the problem.

Flow: Continuously moving the product or information from one operation to the next.

Gemba: A Japanese term for "actual place" or "the place where it happens." In manufacturing, the Gemba is the shop floor, where all value is created.

Gemba Walk: Refers to a walk of the process or "place where it happens" to identify flow, bottlenecks, and waste in a process.

Hypothesis Testing: Assuming a possible explanation to the problem and trying to prove (or, in some contexts, disprove) the assumption.

Inventory: Refers to the raw materials, parts, subassemblies, and finished goods that a business holds to support its operation. These items represent the total amount of capital investment tied up and available to be converted into finished products and sold to customers.

JIT (Just-in-Time): Production methodology characterized by continuous one-piece-at-a-time flow production accomplished according to takt time, the pulling forward of inventory through signals generated by customer demand, and using the absolute minimum resources of man, material, and machines to make only what is needed—when it is needed.

Kaizen: A Japanese term meaning "change for the better" or "incremental improvement." In lean manufacturing it is defined as a team event to identify the root causes of waste and find solutions to eliminate them.

Kitting: A presentation method that groups parts and/or materials. It is a visual management technique used to prevent a worker from starting to build an assembly only to find that parts are missing or not available.

Lean Event: A rapid improvement or kaizen event.

Lean Manufacturing: A Just-in-Time (JIT) manufacturing system.

Material Presentation: A method of introducing material to the line or cell that is oriented in a way to make it easy and effective for the worker.

Non-Value-Added: Any activity that does not add value to the product or service but absorbs resources and increases cost.

PDCA: An acronym for *plan, do, check,* and *act.* It is known as the Deming Cycle. The process is to develop a plan with defined and focused activities, do the activities defined in the plan, check the output from the planned activities, and determine if the outputs meet the planned expectations; if no, take action to correct the problem.

Poka-Yoke: A Japanese term for a mistake-proof device that is designed to stop parts, processes, or procedures from being assembled or used incorrectly.

Point of Use (POU): A location that is within easy reach of workers and their workstations. It eliminates the waste of searching for items.

Process Mapping: A technique of documenting the detailed flow of a part or delivery of a service throughout the required cycle of steps to completion.

Pull System: A method of triggering production by product consumption. When a customer purchases a product, the upstream operations are activated to replenish the item.

Push System: The opposite of a pull system. A supply process is triggered by a sales forecast, which is an assumption about customer demand based on historical data. This methodology creates waste in the form of overproduction and excessive inventory.

Quality: Meeting all expectations and requirements, both stated and unstated, of the customer.

Root Cause Analysis: Eliminating the cause of the problem.

Sequence of Work: The order in which an operator performs a series of repetitive tasks.

Seven Types of Waste: Overproduction, excessive inventory, transportation or movement of material, unnecessary motion, waiting, overprocessing, and defects.

6S Program: Refers to the 5S program of sort, set in order, shine, standardize, and sustain with an additional *S* for *safety.*

Six Sigma: A quality system to achieve a measure of 3.4 defects per million opportunities.

Standard Work: A method for identifying the best practices and documenting them.

Takt Time: The pace at which a production process must work to meet customer demand. Takt time is calculated using the available time divided by the customer demand. *Takt* is a German word for *beat* or *pulse.*

Toyota Production System (TPS): Considered by many to be the benchmark for world class.

Travel Diagram: A drawing of work area layout. It illustrates the movement of people, parts, and equipment to and/or from a workstation or staging location. Travel distances are captured to determine how far they actually travel during a typical workday.

Trial-and-Error: Testing possible solutions until the right one is found.

Value-Added Work: Work activity that changes the fit, form, and function of a product. It is something the customer is willing to pay for.

Value Stream: Refers to all the value-added and non-value-added activities required to design, order, produce, and deliver finished goods into the hands of the customer.

Value Stream Mapping: A method of drawing a diagram that is a graphical representation of every step involved in the information and material flows of a process.

Visual Controls: The tools of visual management such as color coding, charts, Andons, schedule boards, labels, and visual flow production lines or cells. The purpose of these tools is to identify any normal versus abnormal situations.

Visual Management: A system of recognition to tell at a glance if a production process is working in a normal state or an abnormal state.

Waste (Muda): Any activity that absorbs resources and does not add value for the customer. The eight types of waste are (1) overproduction, (2) excessive inventory, (3) transportation, (4) defects, (5) waiting, (6) excessive motion, (7) overprocessing, and (8) underutilized people.

Water spider: A multiskilled, well-trained person who has a routine to replenish the materials and/or parts of a cell. They know all the processes of the production operations thoroughly enough to step into any work position if needed.

WIP: Work in Process, which is inventory that is being processed through an operation.

Index

A

Abuse of coaching, 41, 44
Accident investigation, 20, 25–26
AGVs, *see* Automatically guided vehicles
Air emission, 65
Analog Devices, 12
Aspect analysis, 97
Automatically guided vehicles (AGVs), 56

B

Bacon, Francis, 30
Balanced scorecards (BSC), 12–14
   business process perspective, 13
   customer perspective, 13
   financial perspective, 13–14
   learning and growth perspective, 13
Benchmarking, 14
Boy Scouts, 120
BP, *see* British Petroleum
BP Deep Water Horizon incident, 15
British Petroleum (BP), 111
BSC, *see* Balanced scorecards
Bubble, 119

C

Carlin, George, 101
Case studies, *see* Lean enterprise, case studies in
Challenger Shuttle incident, 119
Change, stress, and innovation, 101–114
   DOT compliance, 102
   EHS, standardized work for, 110–111
      hourly workers, 110
      process safety management programs, 111
      safety performance, menu method of measuring, 111
      SCRAPE program, 110
      top management, question for, 111
   EHS professionals, standardized work for, 111–114
      documentation of standard work, 111
      Excel, 113
      resistance to change, 112
      step-change process management, 114
      visual board, 112
   hybrid approach, 98

Outreach Training Institutes, 102
   post-kaizen checklist, 102, 103
   time studies, DILO, 108–110
      application, 110
      categories established, 109
      employees, 109
      management, 109
      value-added versus non-value-added work, 108
   training kaizen team members in safety and health, 102–108
      elective training modules, 105–108
      mandatory training modules, 105–108
      regions, 106–108
Correction waste, 54–55
Covisint, 92
Current state map, 49
Cycle time, 56

D

DART, *see* Days away and restricted time
Day-in-the-Life-of (DILO) study, 108–110
   application, 110
   categories established, 109
   employees, 109
   management, 109
   value-added versus non-value-added work, 108
Days away and restricted time (DART), 45, 66
Decision making, *see* Planning, decision making, and problem solving
Deming, William Edwards, 3–4, 30, 87
Deming's 14 points, 7–11
   add value to organization by developing your people, 9
   continuously solving root problems drives organizational learning, 10
   long-term philosophy, 7
   right process will produce right results, 7
DHS, *see* U.S. Department of Homeland Security
DILO study, *see* Day-in-the-Life-of study
DOT compliance, 102

E

EHS, *see* Environmental health and safety
Eight Disciplines of Problem Solving, 35–37
Einstein, Albert, 49

143

For Product Safety Concerns and Information please contact our
EU representative GPSR@taylorandfrancis.com Taylor & Francis
Verlag GmbH, Kaufingerstraße 24, 80331 München, Germany